Knowledge and Knowledge Systems:

Learning from the Wonders of the Mind

Eliezer Geisler
Illinois Institute of Technology, USA

IGI PUBLISHING
Hershey • New York

Acquisition Editor:	Kristin Klinger
Senior Managing Editor:	Jennifer Neidig
Managing Editor:	Sara Reed
Development Editor:	Kristin Roth
Typesetter:	Jamie Snavely
Cover Design:	Lisa Tosheff
Printed at:	Yurchak Printing Inc.

Published in the United States of America by
      IGI Publishing (an imprint of IGI Global)
      701 E. Chocolate Avenue
      Hershey PA 17033
      Tel: 717-533-8845
      Fax: 717-533-8661
      E-mail: cust@igi-global.com
      Web site: http://www.igi-global.com

and in the United Kingdom by
      IGI Publishing (an imprint of IGI Global)
      3 Henrietta Street
      Covent Garden
      London WC2E 8LU
      Tel: 44 20 7240 0856
      Fax: 44 20 7379 0609
      Web site: http://www.eurospanonline.com

Library of Congress Cataloging-in-Publication Data

Geisler, Eliezer, 1942-
  Knowledge and knowledge systems : learning from the wonders of the mind / Eliezer Geisler.
    p. cm.
  Summary: "This book explores the process of knowledge formation and how humans generate complex constructs from their sensorial inputs. It also describes how we store and utilize this complex knowledge in organizational and technological systems that allow us to communicate, generate and manage social interactions and functions, providing value-rich content on a cutting-edge issue to researchers, scholars, and practitioners"-- Provided by publisher.
  Includes bibliographical references and index.
  ISBN-13: 978-1-59904-918-2 (hardcover)
  ISBN-13: 978-1-59904-920-5 (ebook)
  1. Knowledge management. 2. Knowledge, Theory of. 3. Knowledge representation (Information theory) 1. Title.
  HD30.2.G445 2008
  658.4'038011--dc22
                         2007022545

British Cataloguing in Publication Data
A Cataloguing in Publication record for this book is available from the British Library.

In loving memory of my parents, Haya and Mordecai (Max),
and to my sister Hannah and her family

Knowledge and Knowledge Systems:

Learning from the Wonders of the Mind

Table of Contents

Section I:
The Structure of Knowledge

Section II:
The Structure of Knowledge

Section III:
Epistemetrics: The Metrics of Knowledge

Foreword

This book is about a journey, the author's intellectual journey in search of an understanding of that strange, ethereal thing we call knowledge. Like all good travel books, it can serve as a guidebook for other travelers on the same journey, whether they have taken it in the past or they embark on it in the future.

Elie Geisler is a knowledgeable and sensitive guide. He brings to the journey the academic disciplines of philosophy, organizational behavior, technology management, and related subjects. He brings years of experience in applying these academic disciplines to practical real-world problems of all kinds of organizations: businesses, government entities, and academic institutions. Most importantly, he brings a lively curiosity about how things work and a tenacious drive to understand

He starts his journey—and our journey—where anyone who is serious about it must start, by defining the geography we will travel together: the structure of knowledge, its components, how the components are formed, and how they are combined into that thing called knowledge. At the end of the first leg of the journey, we and he have reached a common understanding of what knowledge is and how it is formed.

The second leg of our journey explores how knowledge progresses. As human beings grow and mature, as cultures grow and mature, their stocks of knowledge increase (most of the time) and sometimes 'old knowledge' is discredited or superceded by new knowledge. We learn that understanding the progress of knowledge requires a different mindset from understanding the structure of knowledge itself. We see variants of that different mindset that are confusing and non-productive, and finally reach one that describes the progress of knowledge in clear and useful terms.

Knowing what knowledge is, how it is formed, and how it progresses takes us to the threshold of the final part of our journey: understanding how knowledge is applied to better the human condition. We look at individuals, organizations, and society as a whole.

This is a very personal book. It is as much about the way Professor Geisler reached his conclusions as it is about the conclusions themselves. Along the way he draws powerful analogies with biology, the physics of sub-atomic particles (superstring theory), Darwinian evolution, and theology. Fellow travelers will find these analogies fascinating and provocative.

Travelers often go to museums as they travel in order to understand the history behind the current "sights" they see. Professor Geisler has provided us with an intellectual museum of the history of efforts to define and understand knowledge. It is a fascinating place.

The journey toward understanding knowledge with Professor Geisler is not an easy one. Most interesting journeys are not. But it is a feasible one, with many homely examples to ease the road. The trip is well worth the effort. I urge you to take it.

Gerald M. Hoffman
Chicago, Illinois

Preface

In recent years there has been a growing interest in the nature of knowledge, its structure, and its utilization. With the advent and eventual proliferation of computers and databases, knowledge has become a commodity embedded in our tools and in machines we use for everyday needs. We produce today more information (and knowledge) in one day than in all the years since the beginning of human civilization.

Imagine a situation where automobile manufacturers suddenly start producing worldwide more cars daily than in all of the hundred years since automobile production began. The existing infrastructure of roads and highways would certainly collapse—unless we are able to understand how such a flow of vehicles can be manipulated and tamed. Information and knowledge are generated today at an ever growing pace, in quantities and varieties that would have seemed inconceivable to our foreparents not more than two generations ago.

In the twilight years of the twentieth century, scholars and businesses have begun to address the issues of the generation, manipulation, and usages of knowledge. Soon there was a sense of urgency in the creation of knowledge management systems. These models and tools for dealing with the avalanche of knowledge in public and private organizations are still in their infancy—as are the theories and philosophies that underlie their principles and their functioning.

All these events are starting to congeal as critical components of a revised look at an age-old intellectual pursuit of so many scholars over the ages: what is knowledge, how is it structured, and how does it progress. Increasingly, knowledge is permeating every aspect of routine living in developed economies. Its sheer volume and its growing complexity and variability are challenging even the philistines to ponder such hitherto academic queries.

In the last decade of the twentieth century and the dawn of the twenty-first, extensive proliferation of information has made the world a very small place indeed. Progress in telecommunication technology and the Internet has penetrated even the most remote villages in the developing world. A child in any corner of the world has access today to the most extensive collection of human knowledge on any subject and in any depth—just by a click of a personal computer. This truly extraordinary phenomenon is one of the hallmarks of the new century.

As the world becomes smaller and more accessible, information and knowledge grow at an even faster pace. We are becoming a "knowledge economy" (Leonard, 1998; Mokyr, 2002; Stewart, 2001) and a "knowledge society" (Fuller, 2001; Leyderdorff, 2003). Knowledge is now universally considered the engine of economic growth and a strong factor in shaping societal relations. In summary, because of knowledge creation and proliferation, our world and that of our children is dramatically different from the world of our parents.

Ever since my studies in political sciences, business, and philosophy, followed by my relentless effort to understand the way technology behaves in organizations, I have been intrigued by the puzzle of human (and organizational) knowledge. It would, however, be presumptuous to consider the possibility that I can deliver in this book the answers to such age-old queries. I am cognizant and fully embrace the sentiment so aptly articulated by Stephen Jay Gould (2002a): "Instead of suggesting a principled and general solution, I shall ask whether I can specify an operational way…in a manner specific enough to win shared agreement and understanding among readers, but broad enough to avoid the doctrinal quarrels about membership and allegiance" (p. 7). I advance in this book my interpretation of how knowledge is structured and how it progresses. My model builds upon a critique of existing intellectual frameworks, but also offers a conciliatory approach that suggests a meaningful compromise proctored by the new model.

The model I espouse in this book has two distinct characteristics which differentiates it from the current literature. The first is the perspective of knowledge as a construct outside the data-information-knowledge-wisdom continuum. Although this distinction is not an original idea, it nevertheless emerges from my perspective of knowledge as the cumulation of sensorial inputs.

The discussion of knowledge outside the data-information continuum is not merely a minor semantic distinction. Rather, it is a major conceptual differentiation between what constitutes knowledge and the notion of knowledge as simply some outcome from information. As we progress along this continuum, there also emerges the notion of wisdom. In this book, I consider wisdom to be a variant of knowledge, particularly knowledge-in-use. This definition distinguishes the knowledge that is useful and that serves some purpose for the knower.

The second distinct characteristic of knowledge in this book is the perspective of knowledge as the cumulation of sensorial inputs in the human mind. Instead of the evolutionary model in which knowledge develops from lower forms of information, in this book I view knowledge as the clustering of sensorial inputs, or sensations. These are cognitive processes, not the processing of information.

Such a model inevitably leads to conclusions about life, ethics, and the existence of an Almighty—all as a logical consequence from the view of knowledge as mental cumulation of sensorial inputs. Every experience or interaction of the mind with its external world is a different set of sensorial inputs with different modes of clustering. When these are added to the existing pool of knowledge previously formed with sensorial inputs, there is a cumulation effect that produces the arsenal of knowledge we possess. Memory and learning are the creation of patterns of sensorial inputs and their clustering in the mind.

In this book I reformulate the knowledge argument and critically discuss the concepts of qualia and consciousness. Knowledge therefore is a highly personalized event, so that the knower has great difficulties in sharing such cognitive outcomes in his mind. This leads to the conclusion that most of what we call "knowledge" is tacit, hence being to a large extent restricted to the knower. There is only a small amount of what we know that can be shared with others. Despite all our human effort as social beings with language, communication tools and processes, and organizational systems such as knowledge systems, we find it very difficult to freely exchange and to effectively diffuse the knowledge we possess.

The implications for organizations and for their KMSs (knowledge management systems) are far-reaching. KMSs have not been very successful. I discuss in this book the reasons for this phenomenon. In principle, the lack of success of organizational KMSs resides in the inherent inability of individuals in the organization to share what they know. It is not necessarily due to their selfishness or unwillingness to part with what they know, but mainly because they can only transfer and share a very small portion of their knowledge. This is, in part, why organizational incentive packages designed to encourage people to share their knowledge do not seem to be effective.

In the years since my study of the clustering of organizational indicators and the framing by technology managers of organizational phenomena (Geisler, 1979), the field of managerial cognition has grown in a vigorous pace. Fees and Zajac (2006), for example, studied the symbolic management perspective and the organization's framing of strategic changes. They refer to framing processes and cognitive sensemaking as the means by which managers understand and interpret their environment and critical events within them. In Geisler (1979), I conclude that managers are able to frame complex interpretations of organizational phenomena with the use (or clustering) of a small number of different indicators of their organizations' processes and structural dimensions. The emphasis was on the *different* indicators and how they are clustered in the managers' minds.

Fees and Zajac (2006) took this a step further by extending the notion to specific strategic notions of how managers view the functioning of their organization. They arrived at two framing approaches—acquiescence and balancing—which they tested empirically.

The process that describes the creation of sensemaking and the framing of strategic approaches by managers is similar to that of knowledge generation. Managers are clustering different indicators of the patterns of growth of their organization, so as to arrive at a perspective that "makes sense" to them and adequately describes their changing environment.

What this means for organizations is the inevitable conclusion that knowledge is an individualized cognitive phenomenon and that there is a need to radically change the way we communicate and share knowledge in human organizations. In this book I argue for a "shift in *the* paradigm" of knowledge systems in organizations. Currently the focus of both research and practice is on improving the means and methods of transforming "tacit" into "explicit" knowledge, so that we can increase the amount of knowledge shared by organizational members. This approach is also derived from the writings of Japanese scholars such as Ikujiro Nonaka and his colleagues.

A different approach advocated in this book suggests the shifting of emphasis to improving the *communication* and *exchange* mechanisms. Whatever knowledge can be transferred (limited by the nature of knowledge as a cognitive phenomenon embedded in our minds), one has to improve the means by which such knowledge can and will be transferred to others. We should not be attempting to extract more tacit knowledge, but strengthen the transferability of that portion of knowledge that individuals are indeed capable of sharing and diffusing.

In this book I also advance the different model of knowledge: the "neuronal model." This model is a culmination of introspection and the application of several disciplines with which I became familiar throughout my academic career.

This model further advances the notion that sensory elements are clustered by a *knower,* in successive iterations, leading to the creation of knowledge. The progress of knowledge in this model is defined in terms of the concept of *continuous cumulation.* The elements of knowledge are conjoined to form meaningful representation of nature and the knower's reality, and are continuously added to the body of knowledge that exists in the knower's possession and that can be shared by other knowers. This model has far-reaching implications for the design and manipulation of databases and knowledge system, for ethics, for auditing, and for a variety of social and economic phenomena where knowledge is a key ingredient for functions of measurement and assessment.

Unencumbered by previous membership in the intellectual community that investigates epistemological and ontological problems of human thought, I emerge from a more practical field of organization scientists. This is both a plus and a shortcoming, neither of which is abundantly euphoric nor life-threatening. The more practical approach to the structure and progress of knowledge is illustrated (I hesitate to say "justified") in the third part of this book. Examples are drawn from the design and utilization of databases and knowledge management systems, to the vetting of the model in its applications. This is not uniquely a "grounded theory" approach to the analysis (Glaser, 1968), nor is this simply a circuitous mode of claiming that the model is valid because it has practical applications, or that if there are practical applications, the model is valid. Rather, the empirical illustrations attest to the applicability of the model, not necessarily to its superiority over other models of the nature and the progress of knowledge.

This book is written with two audiences in mind. The first is the general public, any person with interest in knowledge and how knowledge management systems and their applications affect our lives. The ubiquitous presence of databases and knowledge systems makes this topic an interesting and exciting area for the modern person to explore. Even more, I believe that our daily encounters with knowledge systems trigger a desire to learn more about them.

The second audience includes academics, information and knowledge professionals, and people involved in the regulation and policymaking of information and knowledge systems. For this audience this book offers a novel approach to the structure of knowledge and its progress. The third part of the book is an empirical showcase of how this structure and progress framework influence the design and manipulation of data and knowledge systems. Upon reading this book, professionals and academics will never approach data and knowledge systems with the same view. Data mining and explorations of knowledge dispersion and transfer will now be predicated on the possibility that the framework advanced in this book is a better explanation of how knowledge is structured and how it develops.

There is something to be said about a style of a book that attempts to advance the state of the art, yet appeals to a broad audience. The style of this book avoids the over-abundance of esoteric terms. Although I had to create some terminology that better explains my ideas, the overall tone of the text is to simply and concisely describe and explain even the more egregious concepts and ideas.

Although in this book I make extensive use of references from the relevant literatures, these sources are mainly employed to give a clearer picture of the prior work in this area. To the knowledgeable reader they serve as a list of readings to consult. To the lay person, the bibliography gives an initial review of what I consider to be the key studies in the area of knowledge, its measurement, and its systems.

A few words about the implications for organizations and the nascent field of knowledge management. The unique model of knowledge advanced in this book indeed has many applications in organizations. It explains not only why KMSs tend to largely be unsuccessful, but also how this situation can be rectified by shifting the paradigm or approach to organizational knowledge management. The model also allows for the development of organizational policies and incentives that recognize the nature of human knowledge and the limitation of its sharing and diffusion, and promote a better way in which people can exchange what they know and benefit from such interactions.

The text aspires to provide a balance between the discussion of theory and the application of the notions and constructs that the theory had advanced. I am a strong believer in Kurt Lewin's pronouncement that there is nothing more practical than a good theory. If we are to truly understand the way in which knowledge is structured and the way in which it grows and progresses, we must be guided by a good theory, and this book offers such a theory.

Stephen Jay Gould (2002b) once argued that theories need both essences and histories. In this book I contemplated the development of theories that attempted to describe knowledge and explain its peculiarities. On their shoulders, as Hawking (2002) has written, I devised an emerging theory (and its applications) that is the crux of this narrative.

References

Fees, P., & Zajac, E. (2006). The symbolic management of strategic change: Sensegiving via framing and decoupling. *Academy of Management Journal, 49*(6), 1173-1193.

Fuller, S. (2001). *Knowledge management foundations.* London: Butterworth-Heinemann.

Geisler, E. (1979). *An empirical study of a proposed system for monitoring organizational change in a federal R&D laboratory.* Unpublished Dissertation, Northwestern University, USA.

Glaser, B., & Strauss, A. (1967). *Discovery of grounded theory: Strategies for qualitative research.* Berlin: Aldine de Gruyter.

Gould, S.J. (2002a). *I have landed: The end of a beginning in national history.* New York: Harmony Books.

Gould, S.J. (2002b). *The structure of evolutionary theory.* Cambridge, MA: Belknap Press.

Hawking, S. (2002). *On the shoulders of giants.* Philadelphia: Running Press.

Leonard, D. (1998). *Wellsprings of knowledge: Building and sustaining the sources of innovation.* Boston: Harvard Business School Press.

Leydersdorff, L. (2003). The construction and globalization of the knowledge-base in inter-human communication systems. *Canadian Journal of Communication, 28*(3), 267-289.

Mokyr, J. (2002). *The gifts of Athena: Historical origins of the knowledge economy.* Princeton, NJ: Princeton University Press.

Stewart, T. (2001). *The wealth of knowledge: Intellectual capital and the twenty-first century.* New York: Doubleday & Company.

Acknowledgment

In the preparation of this book, I was very fortunate to have had inputs and encouragements—as well as criticisms—from many scholars, valuable friends, and from family members. Janet Goranson worked with me from the very inception of this book. She helped me to put it together and was a constant contributor to the intellectual development of my ideas. I am deeply grateful to her.

Kristin Roth, my development editor at IGI Global, and the IGI editorial staff supported this book with guidance and encouragement. I owe them a debt of gratitude.

I am also indebted to my many teachers, colleagues, and students who helped shape my ideas, notions, and views on knowledge. But, in the final analysis, the sole responsibility for the ideas expressed here and the content of this book are mine alone.

I am deeply indebted to my father Max Geisler, who, in his short life, instilled in me the pride, the fervor, and the unabashed delight in the pursuit of knowledge.

To my wife Betsy, for her love and support throughout the making of this book, I am eternally grateful.

Section I

The Structure of Knowledge

Chapter I

Defining Knowledge:
What is Knowledge

What is Knowledge?

In virtually every aspect of our lives, we deal with knowledge. From the moment of birth, we are a sieve through which data is constantly feeding and knowledge is created. We are told that "knowledge is power" and the gateway to prosperity (Allee, 2002; Burke, 1999; Leonard, 1998). Since antiquity, people with inquisitive minds have attempted to define, classify, and measure this illusive notion of knowledge.

It is definitely not an easy task. Philosophers and scholars from a variety of disciplines have failed for many centuries to make this notion amenable to taxonomies and explanation. But, is the topic of knowledge also of interest to the intelligent general reader? The editor of a respected publisher once expressed doubts that such readers would be interested in an abstract notion such as knowledge. Although it may well be illustrated with a variety of examples, the editor exclaimed its focus is such a broad question that it can only be explained with esoteric terminology. The general reader, however curious and educated, would find it uncongenial. I beg to differ. I have much more confidence in the general reader. True, some of the terms in this book are scholarly and perhaps unfamiliar. Terms such as epistemology, although they represent complex phenomena, are not repulsive nor complicated. Epistemology is a branch of philosophy concerned with knowledge. Its base are two Greek words: *episteme* (knowledge) and *logos* (theory). When accompanied by direct and concise narrative, these terms become companions to a delightful understanding of the notions they define. Hawking (2002) has done so. He made physics and astrophysics a popular theme. Gould (2002) made evolutionary theory an exciting topic of general interest. It has been done and I attempt to do so in this book.

So, the question, What is knowledge?, can be answered in three different yet complementary streams. The first would explore *the way in which knowledge is structured*. That is, what are the elements that make it? What is knowledge composed of? Is it data, information, or some other component? The second way would examine *the nature of the dynamics and progress of knowledge*. It would attempt to answer questions such as: How does knowledge progress? How does knowledge accumulate and grow? What are the principles we may discover that explain the patterns of growth and progress? The third stream would examine the *uses of knowledge* in the lives of individuals and how they apply their knowledge to their involvement in the economy and social affairs of their communities and their nations. This stream would focus on the ethics of using knowledge and the means by which knowledge is put to the test of human activities (Bali, 2005; Bock, Smud, Kim, & Lee, 2005; Kankanhalli & Tan, 2005).

Indeed, this third stream has been the focus of attention of scholars for over two millennia. They concentrated on the ethics of knowledge possession and utilization, and on the ways and means in which it serves as a tool for human action (Artigas & Slade, 1999; Cassirer, 1950; Feyerabend, 2000; Kant, 1999). They formulated theories and arguments that explored the role played by knowledge in religion, ethical behavior, and social involvement of individuals and organizations.[1]

But, despite all this intellectual effort, there was very little learned about the *structure* of knowledge and its progress. As the reader will find in the following pages, only very recently has there been some intellectual work toward theories of evolution of knowledge and its more fundamental classification.

For instance, Kant (1999), the influential German philosopher (1724-1804), critically addressed the nature of knowledge. He proposed a structure of knowledge composed of two distinct forms of "knowing" the world. One was *empirical* statements or propositions, which are dependent upon sensory perception. The other was *a priori* propositions (categories). These are fundamentally valid and are not the result of sensorial perception (as Kant called them: "intuitions"). But Kant stopped short of asking: What are the components or basic elements of knowledge? How is knowledge created in the human mind (beyond intuitions and categories)? How does knowledge grow, accumulate, and progress? To Kant's credit (and this is a good indicator of his work being broad and conceptual), he influenced political theorists such as Karl Marx. His philosophical work on knowledge influenced a substantial number of theorists, such as Hegel, Fichte, and Cassirer.

Kant and many other scholars who followed were engaged in this tremendous effort to classify knowledge and to try to understand it as a function of human existence. Although some of my colleagues will certainly disagree with the following statements, this effort remained very broad. The literature, since the age of enlightenment in the seventeenth century, has focused on arguments, counter-arguments, and constructs of knowledge that, although very insightful, were nevertheless at a high level of abstraction (Jacobs, 1999; Meek, 2003). The tasks of defining knowledge by its components and measuring its growth and progress remained woefully unfinished. It is the first two streams that I cover in this book. The initial stream is the *structure* of knowledge, followed by the discussion of the *progress* of knowledge.

It is no coincidence that there is an increased interest in defining knowledge, and that such a definition includes data and information as key ingredients of the answer to what is knowledge. Let me put into perspective the recent developments in the scientific inquiry and applications of knowledge.

Why the Recent Surge in Interest?

Much has been published in the late 1990s about the "information age" and the "information society."[2] Indeed, the twenty-first century began with the effects of the information age spreading to almost every corner of the world and to virtually every aspect of economic and social living. But this phenomenon began almost half a century earlier.

After the Second World War, the invention and proliferation of computers had generated changes in the way industry and government conducted their operations. Initially, in the 1950s and 1960s, computers (or, as later called, "information and telecommunication technology"—ITC) were confined to large scientific, business, and government applications for the purpose of "data crunching." Scientists dealt with massive amounts of data they needed to calculate, and business and government organizations struggled with accounts payable, receivables, payrolls, and inventory control. All of this was confined to what we call "backroom operations," hidden from the public eye but crucial to the routine activities of large organizations. The software at that time was relatively uncomplicated. Languages had emerged with names such as Fortran (science-oriented) and COBOL (business-oriented). Later, other languages began to appear (BASIC, PL1, and IBM's C family of languages).

The impetus for the emergence of the information revolution came in the form of three converging phenomena that began in the 1970s. First, the performance of computing power (hardware) soared, due to the introduction of integrated circuits. This, in turn, led to a continuous relative decline in cost of computing (Moore's Law).[3] The second phenomenon was the invention and rapid proliferation of desktop computing (personal computers, or PCs). Invented in the 1970s, these machines soon appeared in corporate and government offices, thus venturing beyond the "backroom" to the "front offices" of managers in all the functional departments of their organizations. A tremendous boost to the diffusion of PCs and to the ease of their usage was delivered by Microsoft and its high-performance operating system. The third phenomenon was a consequence of the first two, and the advent of the hypertext software which facilitated the proliferation of electronic communication in the form of the Internet and the World Wide Web networks. The phenomenon was an enormous spike in the level of investments in ITC by business and government organizations. In the last fifteen years of the twentieth century, companies invested heavily in information and telecommunication technologies, and made these technologies ubiquitous throughout their organizations. More hardware, better software, and growing budgets for maintenance of systems were expanded to such areas as manufacturing, marketing, and even customer relations.

We sometimes tend to look at the sudden rise and precipitous fall of the Internet-based companies of the 1990s with awe, perhaps even incredibility. How could all this happen in one decade? Yet, this story of "boom and bust" obscures the fact that the basic elements of information technology and the Internet were not the only factors that hastened the information age. As electronic networking became a reality in businesses, government, and private homes, investments in these systems continued to soar. The more we had, the more we used, and the more we wanted of the digital world that now became an integral and indispensable part of our lives.

As a result, business and government organizations were faced with the annoying problem of huge investments in information technologies, at a rate of growth that seemed intractable and unstoppable. Simultaneously, they began to question the benefits that supposedly were being derived from these runaway expenditures.

The Productivity Paradox

Even before the attempt to link investments in information technology to productivity, economists and other scholars pondered over the link between investments in research and development (R&D) and industrial productivity. In general, it was very difficult to show with actual statistics that investments in technology engender corresponding increases in corporate productivity. Although in the mid-1980s there emerged a widespread view that we were entering "the age of productivity" in the U.S. and world economics, the numbers simply did not add up. Productivity is measured as real outputs per worker, where we correlate multiyear investments in information and telecommunication technologies with gains in productivity in the American economy. The results failed to show an increase in productivity that could be attributed to these investments. This phenomenon was termed the "productivity paradox."

If indeed such a paradox exists, there are several explanations for it. Some economists believed that despite massive investments in ITC (which should have boosted plant and office productivity), the costs of running businesses had increased in those years to an extent that they tended to offset gains from ITC investments. Another explanation pointed to the cost and effort in implementation and absorption of ITC into corporate routines—so that gains in productivity are delayed or overshadowed by the "learning curve" effect of the introduction of new technologies. Another explanation focused on the apparent problems in measuring productivity as a consequence of investments in ITC. That is, we lack adequate instruments to measure accurately the link between investments in ITC and the resulting changes in productivity.

From Information Technology to Knowledge Management

The "productivity paradox" had two important outcomes. As the 1990s came to a close, it became abundantly clear to both scholars and managers that investments in information technologies are a fixture in economic life. Regardless of the level of benefits they produce, the trend for the foreseeable future is for more, not less, investments in these technologies. Therefore, if productivity is not the best indicator of the benefits from ITC, we must look elsewhere.

All this led to the second outcome: the shift to a focus on ITC and its benefits as *intellectual assets.* The concept is hardly new, but its application gained vigor in the late 1990s as a strong alternative to explaining the contributions derived from information technologies.[4] The basic idea was the application of the capabilities of current information technologies to collect, assemble, store, manipulate, and diffuse business-related knowledge. The quest thus intensified to understand how knowledge can be harnessed in corporate mission and activities. If the 1990s were the period in which ITC systems had been created and introduced into organizational operations, the twenty-first century would be the age in which we put these systems to use so that human and organizational knowledge can be utilized as an economic asset. This seemed to be a natural extension of the previous period in which investments in ITC had laid the foundation for such a pursuit of the role of knowledge in organizations.

The net result was a massive entry by information and management scholars into the study of knowledge as an organizational and managerial discipline. These scholars took over where philosophers have dwelt for centuries. Their focus was on applications, manipulation, results, and efficiencies. The questions they asked had shifted from "What is knowledge and how does it affect beliefs, behavior, and ethics?" to "How can knowledge become a factor in organizational success?"

This historical account places the development in knowledge management within the context of the key trends in the study of information technologies and their applications. As a nascent effort, the literature on knowledge management shows very little focus on the need to identify or measure the elements of knowledge. Much of the effort is on taxonomical attempts to get a general idea of what the types of knowledge are, mainly in the framework of organizational life.[5]

The Quest for Knowledge as an Asset

What I described above is the current effort to understand human and organizational knowledge as a technological and economic asset. Defined in terms of intellectual *property,* knowledge is thus viewed by current scholars as a variable similar to capital, land, and equipment—as a factor in the production of goods and services. Broader terms such as "The Knowledge Economy" and "The Age of Knowledge" are even stronger indicators of this trend.[6]

Yet, the process by which knowledge does impact the economic behavior of organizations is not adequately understood. The quest remains within the limited region of very broad models of the utilization of knowledge (and information) by economic and social organizations. Knowledge is viewed as a quantity, at best defined in terms of "nuggets," propositions, and predicates. The focus is not on what constitutes knowledge, but on how it flows, how it is absorbed, and how it is implemented to create wealth.

The operational answer to this quest is the knowledge management system (KMS). Designed and established as an organizational creation, such systems are supposed to be the vehicles for knowledge exchange, as well as the physical container for the storage and flow of knowledge. Synnott (1987) argued in the early days of this quest that the information revolution had three phases: hardware (1980-1985), software (1985-1990), and knowledgeware (1990s and beyond). Encapsulated in management systems, knowledge could then be utilized, directed, and put to good use as are other forms of economic factors, such as capital and land. Once hardware and software were in place, the next logical step would be to harness knowledge. This did not happen as well or as fast as was hoped. Knowledge management systems are a far cry from the aims and dreams of their creators. Rubenstein and Geisler (2003), for example, listed several categories of factors that seem to act as barriers to the successful performance of these systems. They include factors inherent in the system itself (focus, search capabilities), human factors (such as fear and unwillingness of organizational members to use them), and organizational factors (how the system is implemented).

Thus, the failure to create and implement truly workable knowledge management systems has led to an even more vigorous focus on the effectiveness and applications of these systems. This, of course, is at the expense of an effort to better measure the elements of knowledge and to improve our understanding of how knowledge is structured. To an extent, the previous effort at metrics of information and information theory has not been fully translated into a similar program aimed at knowledge. It is lamentable that such a natural progression has not occurred. Knowledge, in the context of managerial and organizational studies, has remained merely a variable in the search for performance and effectiveness. I return to this topic in Chapter XIII to revisit the applications of knowledge systems as seen through the prism of the model I advance in this book of how knowledge is structured and how it progresses.

Bridging Philosophy and Management

The quest for the definition of knowledge has now assumed two seemingly independent routes. Traditional epistemology (the study of knowledge) is a mature area of philosophical inquiry. Simultaneously, as I described above, management scholars are increasingly addressing the topic of knowledge and its role in organizations. It seems to me that there is a need to bridge the two streams of intellectual pursuit. What is it that we know about knowledge from philosophers, and how can we join this with what we know about knowledge from management scholarship?

A review of the relevant literature will show that there has been little exchange between the two streams. Partly because of disciplinary isolationism and the different objectives of their pursuit, the two streams lack workable bridges that would allow them to share ideas and methodologies.[7] Some approaches to knowledge definition are at the margins of the two streams. For example, research on epistemic value, information theory, and information ethics may produce results that would be useful to scholars who study knowledge management systems.[8]

In this book I embark on the path of trying to bridge the two streams. If, I pondered, we can advance a plausible model of the structure and growth of knowledge (as an outflow of current philosophical models), and make the connection of this model to knowledge systems—the claim to relevancy will be answered. Thus, the definition of knowledge, as a start, will be based on prior art in philosophy, but will also be developed to frame a model with tremendous implications for the design and application of knowledge-based systems. The first step will require a review of the widely used typology of knowledge and its constituents.

The Great Taxonomy:
Data Information and Knowledge

Anytime we approach the construct of knowledge, it will inevitably be linked to information or described in terms of information and its attributes. The most prevalent taxonomy is the triad: data, information, and knowledge. Presented as a hierarchy of the components of information systems, this taxonomy defines *data* as 'streams of raw facts representing events occurring in organizations or the physical environment before they have been organized and arranged into a form that people can understand and use."[9] I prefer the following: "Data are observations on the physical or human world coded in a form that can be stored, manipulated, transferred, and shared."

However we define *data,* these are things that describe the world in a form that people and machines can both understand and use. The utilitarian approach will distinguish "just raw facts, or observations" from *data,* which is usable within a well-defined context of human or machine capabilities. If, for example, a tree falls in the forest, this fact by itself does not constitute datum, unless it can both be understood and *potentially useful.* This means that by force of the definition, a data point or unit (datum) can only be so defined in relation to an entity external to the datum: a biological being or a machine, capable of understanding it and possibly also using it.[10]

Information is usually defined as "data that have been shaped into a form that is meaningful and useful to human beings" (Laudon & Laudon, 2002). This definition contains two parts. The first describes the flow or volume of information that moves through a channel. This is flow without inherent meaning, simply measuring the capacity of a channel to move volumes of data, also known as "syntactic" information. The second describes an inherent meaning (semantic) to the flow of data.

These definitions are important inasmuch as they guided the research that follows. For example, Shannon's theory of information initiated a stream of research into the flow of information, transmission channels, and the mathematical representation of communication systems between humans, machines to machines, and humans-to-machines.[11] For Shannon, the unit of information has the "bit" (a zero or a 1) in the digital world of computers. He shows how we can compute channel capacity to transmit bits/second thus providing a quantitative model of communication and information transmittal.

As soon as the definition includes conceptual meaning in the information, the result is a dependence on the interpretation of information and the creation of "meaning"—to and by a knower who is able to make such an interpretation. In turn, this ability requires some criteria for analysis of the information, in order to make sense of it and extract its meaning. Thus, the meaning is not embedded in the information, but in the receiver (or knower), who has the analytical or rational tools to extract meaning.

Knowledge is often defined as "actionable information" (Nonaka & Takeuchi, 1995). Here there is a more extensive reliance on the external entity who naturally must be able to extract meaning, but also must have the capability of integrating such information into a plan of action—that is, being able to act on the basis of the information received. This definition is concise as it is very broad. It does not include a clear reference to what action is, nor how information would be used in action-taking by the knower.[12]

What then do we know about knowledge and its unit? From the taxonomy of data-information-knowledge, we have learned very little. This taxonomy fails to offer a robust hierarchy of complexity or a tractable flow from the elemental to the compound. The unique difference that separates data from information and from knowledge is the reference to an external entity and to the potential of what we are categorizing to perform a given function for the external entity (the knower). To summarize, data are raw items describing reality in a form that can be understood. Once we add "meaning" to this definition, we are now in the realm of information, whereas when we add a tenuous relation to the ability to use it within a given scheme of action, we are now describing knowledge.

There is a tremendous gap between Claude Shannon's definition of bits, and the trinity of definitions from data to information to knowledge. The over-reliance on the external entity brings with it too many variables to consider. For example, who is the entity who will make information actionable? How would such a transformation occur? Under what circumstances, constraints, and conditions will there be another round of interpretation and extraction of meaning as we move from the level of information to that of knowledge? Are these levels hierarchical? How we long for the simplicity of Shannon's

Theory where information resembles a physical quantity amenable to mathematical measurement.[13]

"The Big Rift": Severing the Tie Between Information and Knowledge

The current literature conceives of knowledge as part of a continuum of data-information-knowledge-wisdom. This notion has various interpretations and definitions, but consistently positions the ontology of knowledge as a 'transformation" or as a more complex conception of information (Lueg, 2001; Tsoukas & Vlachiminou, 2001). Kakabadse, Kakabadse, and Kouzmin (2003) reviewed the literature and offered the following definitions: "Data reports observations or facts out of context that are, therefore, not directly meaningful" (p. 77). Information is largely defined as "placing data within some meaningful content, often in the form of a message" (p. 77). By extension, knowledge is defined as "information put to productive use." Thus, when one acts upon information, extracts value from it, and makes it useful for a given end, knowledge is generated. Finally, "through action and reflection one may also gain wisdom. Knowing how to use information in any given context requires wisdom" (p. 77).

These and similar theoretical and empirical definitions of knowledge confine the notion to the analytical and conceptual space of information. It also resorts to an external anchor, such as the utilization of information, as a key dimension of what constitutes knowledge (Kankanhalli & Tan, 2005; Zack, 1999). The outcome is that knowledge is simply the processing of information into a form that produces value to a user and can be utilized for a purpose (Guah & Currie, 2004; Xirogiannis, Glykas, & Staikouras, 2004).

I reject this constrained definition on two grounds. The first is the lack of a clear boundary between the conceptions of information and knowledge. Where does information end and knowledge begin? Even when construed as distinct stages in the flow or chain (from data to wisdom), these stages lack empirical boundaries that distinguish between what constitutes information and the subsequent notion or stage of knowledge (Hunt, 2003).

The criterion of "productive use" or utilization towards a given purpose fails to sufficiently define knowledge as distinct from information. When we define knowledge as a variant of "useful information," this exercise in terminology does not make an advance towards a distinct concept. Information and useful information are similar definitions of the same notion.

Secondly, I reject the definitional link between information and knowledge because it prevents knowledge from being defined as an independent entity, with its own ontological integrity. As a recombinant version of information, knowledge is not conceived of as a distinct notion. The problem is exacerbated because the literature on knowledge management considers knowledge to be a reified concept and possessing a "stand-alone" format (Day, 2005; Muthusamy & Palanisamy, 2004; Reisman & Xu, 1992; Zack, 1999). Although the literature on knowledge management valiantly attempts to disengage itself from the information systems and information research literature, it is nevertheless bounded by the information-laden definition, hence by the methodology of this more established area of research (Hendricks & Vriens, 1999; Pritchard, Hull, Chumer, & Willmott, 2000). When knowledge is viewed as intellectual "assets" or "capital," there is a tendency to reify it as a distinctly measurable entity with organizational antecedents and strategic implications (Bontis, 2001; Glazer, 1998).

Thus we find ourselves inexorably immersed in the claws of the ontological conception of data and information. Yet, we fail to demonstrate where knowledge begins and what independent form it assumes within the information-bound perspectives. Finally, how can the nascent literature on knowledge and knowledge management address the metrics of knowledge without being shackled to the information systems literature, its methodology, and its research topics (Alavi & Leidner, 2001; Earl, 2001; Grover & Devenport, 2001; Martin, 2004)?

The answer to exercising a "rift" or disengagement from the information framework is to assume a radically different perspective to the ontology of knowledge. In the current conception of knowledge, the notion of information—extended to knowledge—is accepted as it was devised and developed by information scientists. There is little, if any, conceptual modification or novel contributions to the notion of information knowledge, except for minor changes to the definition.

The model of the structure and generation of knowledge proposed in this book contends that the starting point of a flow in which knowledge and information participates is knowledge, not data or information. What we know originates in the human mind, where sensorial inputs are clustered, and where perceptible distortions are conjoined to form the basic units of knowledge.

The cognitive processes of the mind in which sensorial inputs (such as a taste, touch, or sound) from our five senses are clustered into knowledge are entirely different from the current conception of information being transformed into knowledge. Rather, humans possess the ability (albeit not always the will) to share, transfer, and transcribe their knowledge in a form usable to others. This transcribed and shared form of knowledge may be termed "information," but the origin of its appearance and ontology is the knowledge clustered in the mind from inputs it derives from the five senses. I do not define such inputs as "data" because, in my opinion, they are not what we commonly defined as data, and also because of the bias inherent in the history of data processing and information systems.

By the current definitions "data" are representations of "facts" or "ideas," whereas sensorial inputs are a very crude form of human cognitive manipulation of inputs from its internal and external environments. The mental conjoining of a flicker of light, a sound, and a sharp momentary pain are different from a "fact" that there are six cars on the road or that freedom is a right of all people.

To grasp these facts and to make sense of them and the "information" they form (in the data → information flow), there is a need for certain structural foundations that will allow such processing. This has been the core of the philosophical and later informational search for such a process, architecture, and categories of the mind (Dalkir, 2005; Kant, 1999; Tsoukas, 2005; Vail, 1999).

These attempts to explain how data converge into information and into knowledge have been largely unsuccessful (Perry, 2005). When the approach is reversed and knowledge is viewed as the initial ontology to be generated by the human mind, then the constructs of data and information become artificial notions that are contingent upon the processes by which humans diffuse, share, and exchange knowledge. Data and information in the model proposed in this book are the reification of the knowledge being transferred from one individual to another (and by extension from individual to an organized knowledge system and between such systems).

The direction of a flow is now reversed. Knowledge is the origin of cognition, formed as a clustering of sensorial inputs (see Chapters IV and V). Once such clustering forms what we term "knowledge," and it is amenable to transfer and to sharing outside the mind, we may now define whatever is transferred as "information."

To transfer "facts" or "information" that there are six cars on the road requires the foundational knowledge of what constitutes cars, road, the number six, and any implications such "facts" or "information" are conveying. In the human mind the only way to form knowledge—hence to be able to absorb such external "information"—is by converting sensorial signals into clusters, which then form the "nuggets" or elemental units of knowledge.

Although the existence of a "flow" of sequential events (such as data-information-wisdom) is an attractive construct, once the direction of this flow is reversed and *knowledge* is the initial construct to be generated, there is no need for a flow of notions. Whether such a flow obeys criteria of complexity (from the simple to the complex) or temporal distinction (from the past to the present), such a flow is a superfluous and irrelevant explanation of the transfer and sharing of human knowledge.

Tacit and Explicit Knowledge

Ever since Polanyi (1966) there has been a widespread acceptance of the distinction between "tacit" or subjective knowledge, and "explicit" or objective knowledge. Nonaka and Takeuchi (1995) extended Polanyi's typology. They defined tacit knowledge as a mode of experience, being simultaneous and analog (practical), whereas they defined explicit knowledge as rational, sequential (there and then), and digital (theoretical).

Michael Polanyi (1891-1976) has argued that humans "know more than they can tell," thus engendering a challenge to successive researchers to identify processes, means, and ways in which we can exercise "knowledge conversion"—from tacit to explicit. The differences between tacit and explicit knowledge (what we know and what we are able to transcribe and to share) became the fundamental content of many models attempting to describe such transcription or conversion (Chua, 2002; Earl, 2001; Eddington et al., 2004; Geisler, 2006). Nonaka and Takeuchi (1995) suggested four modes of conversion: *socialization* (from tacit to tacit), *externalization* (from tacit to explicit), *combination* (from explicit to explicit), and *internalization* (from explicit to tacit). Chua (2002) proposed a taxonomy of organizational knowledge in which the classification scheme entails individual and collective knowledge, and these are further classified into tacit and explicit knowledge.

All knowledge is tacit, the result of the clustering of sensorial inputs as they are cumulatively created in the mind. Hence, the distinction between tacit and explicit knowledge is at best an artificial differentiation. The terms "tacit" and "explicit" simply denote a temporary location of knowledge once it is defined, rather than describing a different ontology.

If knowledge is still in the form of clustered sensorial inputs, encloistered in the human mind, then the distinction is irrelevant, because at this point there is no apparent means or mode to access this knowledge. If, however, the knowledge has been shared in the form of nuggets, for example, then it is purely a matter of existing in the accessible universe of human interactions and communication. Thus, whatever is or is not encloistered or embedded in the human mind is irrelevant, immaterial, and inconsequential to shared knowledge. We now need to better define the clustering of knowledge in the mind on the elemental unit of knowledge and to adequately describe the means by which knowledge is shared among minds. Perhaps we may even arrive at the conclusion that whatever is clustered in the mind is indeed "knowledge" and whatever is shared is not knowledge, to be termed "information" or some other connotation.

"Tacit" knowledge may at best refer to the potential of the human mind to generate knowledge from sensorial signals, and to the repository of such clustered formations in the human mind over the person's lifetime. This may mean that the longer the person is in existence and the mind operates, the more knowledge has been generated and cumulatively deposited in the individual's knowledge base in the mind. This does not take into consideration the quality and other attributes of the content of this knowledge base. One must also take into account the levels of attrition and loss of content due to the aging of the brain and other detrimental biological factors of decay.

Such "tacit" knowledge is only meaningful when it can be measured and useful when it can be accessed and shared. Much of the intellectual effort of scholars and practitioners in the areas of knowledge and knowledge management has been devoted to improving the flow from "tacit" to "explicit" (Alavi & Leidner, 2001; Bock & Kim, 2002; Davenport & Prusak, 1998; Lee & Choi, 2003; Rothberg & Ericson, 2004; Sharp, 2003). On the whole this effort produced puny results, leading to several crises in knowledge management and in the continuing failure of many organizational knowledge systems to perform at the level promised by the discipline. As I describe in Part IV of this book, individuals in organizations are reluctant and even averse to sharing what they know.

Michael Polanyi was correct when he asserted that people know much more than they share, but for the wrong reason. We fail to share more than a small portion of what we know not because we refuse to do so (for personal, organizational, or other reasons), but because we are, by design, unable to do more than we have been doing. Hence, attempts to "improve" such a phenomenon of transfer and sharing of knowledge are bound to fail.

Where Knowledge Resides

The effort to define what we know and what we understand by the notion of "knowledge" also leads to questions about the locus of knowledge. Who has it and where does it reside? The notion of the knower will be amply discussed in this book. This notion has two venues: the human cognition in the mind and machines hosting knowledge.

The appearance of computers in the second half of the twentieth century introduced a new dimension to our view of candidates for hosting knowledge. In addition to their ability to receive and process information, these machines engendered a new area of study—human-machine interaction—that is raising many complex questions. Among these are: How well can machines perform as information (or knowledge) processors when compared with the human mind? How similar are these machines to a human mind? How do humans interact with these machines?

In 1936 Alan Turing proposed an effective method of computation known as the "Turing Machine." Independently, Alonzo Church (1941) had offered a similar conjecture. A Turing Machine is a notion of a computing device or method able to compute any recursive functions.[14] The impact of this notion of machine capability has been substantial. It led to a growing appreciation of the power of machines to harness and to process information, hence knowledge. A natural development was the suggestion that the human mind is essentially a Turing Machine. Conversely, with the adequate programming that emulates human mental processes, machines (computers) could eventually think. This concept is known as "machine functionalism."

The human mind may thus be modeled as a machine. Mental states are viewed as "automatic formal systems" whose outcomes are continuously interpreted as a system of relationships. Knowledge is produced as a result of these computations that define mental state at the level of calculations—similar to what a Turing Machine would be doing.

The computer-model of the human mind was extended to the area of exploration known as *artificial intelligence* (AI). Since the early 1980s, a considerable effort in both financing and intellectual resources has been expended in developing AI devices, components, and models. Previously known as "expert systems," these were computer programs aimed at performing complex tasks, to the extent that meaning, concept formation, and understanding could be achieved. In medicine, for example, expert systems, such as the pioneering *Mycin,* were designed to provide a diagnosis of a disease based on symptoms, medical history of the patient, and the results from relevant tests performed on the patient.[15]

Turing himself had suggested a test by which a machine could be defined as possessing humanlike intelligence. The machine would be interviewed by an educated human being (without direct physical contact between human and machine), and if the interviewer is unable to tell whether the entity on the other side is human or machine, the machine passes the test of intelligence.[16]

The basic notion of this effort in extending cognition to machines is that "intelligent beings are semantic engineers—in other words, automatic formal systems with interpretations under which they consistently make sense' (Beedle, 1998, p. 243). But, although AI systems have become more sophisticated, there is still a very wide gap between their performance and human intelligence capabilities. In over three decades of the existence of expert systems, those that survived are only used for limited tasks rather than in a broad capacity as sources of knowledge. In particular, this applies to such systems as programs that contain and process knowledge. In effect they contain little knowledge, as defined here. None has passed the "Turing test," nor have computers achieved a level of thought comparable to the human mind. More on these machines, expert systems, and artificial intelligence is in Part IV of this book, where I consider the relation and impact of the structure and progress of knowledge on data and knowledge bases and systems.

How Much Do We Really Know About Knowledge?

The effort described above did little to bring together the stream of scholarship by philosophers engaged in epistemology with those in the nascent field of knowledge management. Both streams adopted the level of propositions, language, concepts, and predicates. Epistemologists explore the truth of these propositions, their value, and their role in ethics—among other goals. Knowledge management is concerned with usage and effectiveness of the systems of knowledge. Neither system ventured to the level of the elemental structure of knowledge. In the allegory of physics, they remained in Newtonian physics, unwilling or unable to delve into sub-atomic explorations and the realm of quantum mechanics.

The issue of definitions, as elaborated in the trinity of classifying data to information to knowledge, is a practice largely favored by management scholars and practitioners. The definitions lead to a better operationalization of concepts, processes, and practices. Therefore, the focus on definitions below the traditional level of higher discourse by epistemologists can be credited for the advent of management scholarship and its pursuit of data and information as components of organizational analysis. Such emphasis was, and continues to be, guided by organizational variables of concern to the extent that there are few, if any, incentives to explore the basic components of knowledge—so long as the existing definitions adequately support models and notions of organizational performance and similar analytical pursuits.[17]

In the current state of affairs, we find these streams of research that are centered around distinct objectives with inconsequential confluence or cross-fertilization. The knowledge management "movement" is concerned with its managerial and organizational issues in pursuit of applications, utility, and the exploration of the systemic attributes of knowledge. It is also overly concerned with transforming "tacit" knowledge into "explicit" knowledge. This stream seems bent on inventing its own notions, research questions, hypotheses, and terminology. Thus far the harvest from this endeavor is relatively puny. We know somewhat more about knowledge systems, but little, if anything, more about knowledge itself: its structure and its progress.

What Else Needs to be Known About Knowledge

I am obliged, in a manner similar to those engaged in knowledge management, to create some generic terms and to define or redefine basic notions of knowledge structure and its dynamics. From epistemology we learned of knowledge as a rational intercourse, conversant and exchanged in language, and argued within human experience and the ontological aspects of what is known. Combined with outcomes from the management quarters, the result is a discourse of knowledge at high levels of consideration, and as a utilitarian "thing" employed by the rational mind and by organizations that such a mind can create.

What we do not know at this juncture, and what needs to be known about knowledge, can be summarized in three items. First, we need to know how knowledge is structured beyond the trinity of hierarchical definitions of data, information, and knowledge. We need to know how knowledge is created, what its basic components are, and if we can regard it in such a design as an ontological unit, rather than an assembly of lower-level components. Can knowledge be considered an entity with its *own* structure and characteristics, or are we merely reifying an agglomeration of, say, items of useful information? This question has been amply discussed by epistemologists.[18] What we need to know is centered around the problem of the structure of knowledge, assuming that it is indeed ontologically viable.

Second, we need to know how knowledge progresses and how it grows in size and magnitude. If knowledge is indeed ontologically acceptable as a unit of analysis by whatever taxonomy is applied, size or volume would be one of its characteristics. This would mean not only the actual growth of the stock of knowledge by individuals and their social groupings, but also the diffusion mechanisms by which knowledge is exchanged and transmitted among individuals.[19] Finally, we need to know how a modified model that links structure and progress impacts databases and knowledge systems. In fact, these three items are the three basic tomes of this book.

Why is this important and what might be the contributions of this book to gaining such an improvement in the stock of what we already know? In our daily lives we are surrounded by databases and knowledge systems. They not only contain much information and knowledge about us—who we are and what we do—but they also serve as the basis for our actions. We rely on these systems to make inferences, to interpret our surroundings and the forces that confront us, and to make judgments on what we should and can do.

In the latter aspect of knowledge utilization, we have gained many insights from the work of such researchers as Leon Festinger on cognitive dissonance, and Amos Tversky and Daniel Kahneman on decision making under uncertainty, to list only a few.[20] The psychology of using knowledge to make decisions is better illuminated by these scholars, and we are hence better equipped to understand how knowledge is stored, diffused, and manipulated.

Although we have learned a substantial amount on knowledge and its utilization, there is ample room for a model or theory that links structure *and* progress. I am not referring to a unifying theory (a "theory of the whole") as physicists prefer to call it, but to a modest attempt to advance a model that links the basic structure of knowledge with a corresponding perspective of how such knowledge progresses and grows.

In the following pages the first section of this book reviews the existing theories and models of knowledge. This is then followed by a description of the model I propose for the structure of knowledge. The second section offers the model as it relates to the progress of knowledge. The third section describes the area of "epistemetrics," and the fourth section of the book addresses the impact of the model on databases and knowledge systems, and how we utilize them to interpret our reality and to act upon it.

References

Adams, F. (2003). The informational turn in philosophy. *Minds and Machines, 13*(4), 471-501.

Alai, M. (2004). Scientific discovery and realism. *Minds and Machines, 14*(2), 21-42.

Alavi, M., & Leidner, D. (2001). Knowledge management and knowledge management systems: Conceptual foundations and research issues. *MIS Quarterly, 25*(1), 107-136.

Arrow, K. (1994). Methodological individualism and social knowledge. *American Economic Review, 84*(2), 1-9.

Artigas, M., & Slade, I. (Eds.). (1999). *The ethical nature of Karl Popper's theory of knowledge.* Bern, Switzerland: Peter Lang.

Audi, R. (2002). *The architecture of reason: The structure and substance of rationality.* New York: Oxford University Press.

Bali, R. (2005). *Clinical knowledge management: Opportunities and challenges.* Hershey, PA: Idea Group.

Beedle, A. (1998). Sixteen years of artificial intelligence: Mind design and mind design II. *Philosophical Psychology, 11*(2), 243-251.

Bock, G., Smud, R., Kim, Y., & Lee, J. (2005). Behavioral intention formation in knowledge sharing: Examining the roles of extrinsic motivators, social psychological forces, and organizational climate. *MIS Quarterly, 29*(2), 87-111.

Bontis, N. (2001). Assessing knowledge assets: A review of the models used to measure intellectual capital. *International Journal of Management Review, 3*(1), 41-60.

Cassirer, E. (1950). *Problem of knowledge: Philosophy, science, and history since Hegel.* New Haven, CT: Yale University Press.

Chua, A. (2002). Taxonomy of organizational knowledge. *Singapore Management Review, 24*(2), 69-76.

Church, A. (1941). *The calculi of Lambda-conversion.* Princeton, NJ: Princeton University Press.

Copeland, B., & Proudfoot, D. (1999). The legacy of Alan Turing. *Mind, 108*(3), 187-195.

Cover, T., & Thomas, J. (1991). *Elements of information theory.* New York: Wiley Interscience.

Cross, C. (2001). The paradox of the knower without epistemic closure. *Mind, 110*(438), 319-333.

Dalkir, K. (2005). *Knowledge management in theory and practice.* New York: Elsevier.

Davenport, T., & Prusak, L. (1998). *Working Knowledge.* Boston: Harvard Business School Press.

Day, R. (2005). Clearing up implicit knowledge: Implications for knowledge management, information science, psychology, and social epistemology. *Journal of the American Society for Information Science and Technology, 56*(6), 630-636.

Dertouzos, M. (1997). *What will be: How the new world of information will change our lives.* New York: Harper Collins.

Earl, M. (2001). Knowledge management strategies: Toward a taxonomy. *Journal of Management Information Systems, 18*(1), 215-234.

Eddington, T., & Choi, B. et al. (2004). Adopting ontology to facilitate knowledge sharing. *Communications of the ACM, 47*(11), 85-90.

Fallis, D. (2004). Epistemic value theory and information ethics. *Minds and Machines, 14*(2), 101-117.

Feyerabend, P. (2000). *Against method.* New York: Verso New Left Books.

Fuller, S. (2001). *Knowledge management foundations.* London: Butterworth-Heinemann.

Geisler, E. (2006). A taxonomy and proposed codification of knowledge and knowledge systems in organizations. *Knowledge and Process Management, 13*(4), 285-296.

Geisler, E., Prabhaker, P., & Nayar, M. (2003, July). Information integrity: An emerging field and the state of knowledge. *Proceedings of the PIC-MET International Conference on the Management of Engineering and Technology,* Portland, OR.

Geisler, E., & Ritter, B. (2003). Differences in additive complexity between biological evolution and the progress of human knowledge. *Emergence, 5*(2), 42-55.

Glaser, B., & Strauss, A. (1967). *Discovery of grounded theory: Strategies for qualitative research.* Berlin: Aldine de Gruyter.

Goldman, A. (1999). *Knowledge in a social world.* New York: Oxford University Press.

Gould, S.J. (2002). *The structure of evolutionary theory.* Cambridge, MA: Belknap Press.

Gould, S.J. (2002a). *I have landed: The end of a beginning in national history.* New York: Harmony Books.

Grover, V., & Davenport, T. (2001). General perspectives on knowledge management: Fostering a research agenda. *Journal of Management Information Systems, 18*(1), 5-21.

Guah, M., & Currie, W. (2004). Factors affecting IT-based knowledge management strategy in UK healthcare system. *Journal of Information & Knowledge Management, 3*(4), 279-289.

Hawking, S. (2002). *On the shoulders of giants.* Philadelphia: Running Press.

Hendricks, P., & Vriens, D. (1999). Knowledge-based systems and knowledge management. *Information & Management, 35*(3), 113-125.

Hunt, D. (2003). The concept of knowledge and how to measure it. *Journal of Intellectual Capital, 4*(1), 100-113.

Huysman, M., & Wulf, V. (2004). *Social capital and information technology.* Cambridge, MA: MIT Press.

Jacobs, J. (1999). *Depravity of knowledge: The protestant reformation and the disengagement of knowledge from virtue in modern philosophy.* Aldershot, UK: Ashgate.

Kahneman, D., Slovic, P., & Tversky, A. (Eds.). (1982). *Judgment under uncertainty: Heuristics and biases.* Cambridge, UK: Cambridge University Press.

Kakabadse, N., Kakabadse, A., & Kouzmin, A. (2003). Reviewing the knowledge management literature: Towards a taxonomy. *Journal of Knowledge Management, 7*(4), 75-92.

Kankanhalli, A., & Tan, B. (2005). Knowledge management metrics: A review and direction for future research. *International Journal of Knowledge Management, 1*(2), 20-32.

Kant, E. (1999). *Critique of pure reason.* Cambridge, UK: Cambridge University Press.

Kurzweil, R. (2000). *The age of spiritual machines: When computers exceed human intelligence.* New York: Penguin Books.

Laudon, K., & Laudon, J. (2002). *Essentials of management information systems.* Upper Saddle River, NJ: Prentice Hall.

Lee, H., & Choi, B. (2003). Knowledge management enablers, process, and organizational performance: An integrative view and empirical examination. *Journal of Management Information Systems, 20*(1), 179-228.

Lueg, C. (2001). Information, knowledge, and networked minds. *Journal of Knowledge Management, 5*(2), 151-160.

Luftman, J. (2003). *Managing the information technology resource: Leadership in the information age.* Upper Saddle River, NJ: Prentice Hall.

Martin, W. (2004). Demonstrating knowledge value: A broader perspective on metrics. *Journal of Intellectual Capital, 5*(1), 77-91.

McKenzie, D. (2001). *Mechanizing proof: Computing, risk, and trust.* Cambridge, MA: MIT Press.

Meek, E.L. (2003). *Longing to know: The philosophy of knowledge for ordinary people.* Wheaton, IL: Brazos Press.

Muthusamy, S., & Palanisamy, R. (2004). Leveraging cognition for competitive advantage: A knowledge-based strategy process. *Journal of Information and Knowledge Management, 3*(3), 259-272.

Nirenberg, S., Carbonell, J., Tomita, M., & Goodman, K. (2000). *Machine translation: A knowledge-based approach.* New York: Morgan Kauffman.

Nonaka, I., & Takeuchi, H. (1995). *The knowledge-creating company: How Japanese companies create the dynamics of innovation.* Oxford, UK: Oxford University Press.

Parrini, P. (1998). *Knowledge and reality: An essay in positive philosophy.* Boston: Kluwer Academic.

Perry, I. (2005). Knowledge as process, not data: The role of process-based systems in developing organizational knowledge and behavior. *International Journal of Healthcare Technology Management, 6*(4/5/6), 420-430.

Polanyi, M. (1966). *The tacit dimension.* New York: Anchor Day Books.

Post, G., & Anderson, D. (2002). *Management information systems: Solving business problems with information technology.* New York: McGraw-Hill.

Pritchard, C., Hull, R., Chumer, M., & Willmott, H. (Eds.). (2000). *Managing knowledge: Critical investigations of work and language.* London: McMillan.

Reisman, A., & Xu, X. (1992). On stages of knowledge growth in the management sciences. *IEEE Transactions on Engineering Management, 39*(2), 119-128.

Rothberg, H., & Ericson, G. (2004). *From knowledge to intelligence.* Oxford, UK: Butterworth Heinemann.

Rubenstein, A.H., & Geisler, E. (2003). *Installing and managing workable knowledge management systems.* Westport, CT: Praeger.

Sharp, D. (2003). Knowledge management today: Challenges and opportunities. *Information Systems Management, 20*(2), 32-37.

Skyrme, D. (2001). *Capitalizing on knowledge: From e-business to k-business.* London: Butterworth-Heinemann.

Synnott, W. (1987). *The information weapon.* New York: John Wiley & Sons.

Tissen, R., Andriessen, D., Lopez, F., & DePrez, F. (2000). *The knowledge dividend: Creating high-performance companies through value-based knowledge management.* Upper Saddle River, NJ: Prentice Hall.

Tsoukas, H. (2005). *Complex knowledge: Studies in organizational epistemology.* New York: Oxford University Press.

Tsoukas, H., & Vladimirou, E. (2001). What is organizational knowledge? *Journal of Management Studies, 38*(7), 973-993.

Vail, E. (1999). Mapping organizational knowledge. *Knowledge Management Review, 8*(3), 10-15.

Von Krogh, G., Ichijo, K., & Nonaka, I. (2000). *Enabling knowledge creation.* New York: Oxford University Press.

Xirogiannis, G., Glykas, M., & Staikouras, C. (2004). Fuzzy cognitive maps as a back end to knowledge-based systems in geographically dispersed financial organizations. *Knowledge and Process Management, 11*(2), 137-154.

Young, R. (2004). Wittgenstein's tractatus project as philosophy of information. *Minds and Machines, 14*(2), 119-132.

Zack, M. (1999). Managing codified knowledge. *Sloan Management Review, 40*(4), 45-58.

Zuniga, G. (2001). Ontology: Its transformation from philosophy to information systems. *Proceedings of the International Conference on Formal Ontology in Information Systems,* Ogunquit, ME.

Endnotes

[1]   It is not my intention to clutter this narrative with samples of philosophical writings. But until very recently, those scholars engaged in the discipline of philosophy were the standard-bearers of the intellectual pursuit of knowledge. Readers may wish to further consult the work of several other scholars not listed here, such as Baruch Spinoza, Rene Descartes, John Locke, and, more recently, Ludwig Witgenstein, Willard Van Orman Quine, and John Austin.

[2]   See, for example, Dertouzas (1997), Geisler, Prabhaker, and Nayar (2003), McKenzie (2001), Nonaka and Takeuchi (1995), Skyrme (2001), and Tissen, Andriessen, Lopez, and DePrez (2000).

[3]   "Moore's Law" is named after Gordon Moore, chairman emeritus of Intel Corporation. He observed, then argued, that the number of transistors in each computer chip doubles every 18-24 months, so that the computing power also doubles. Hence, even if prices of computers remain unchanged, they sharply fall relative to the number of computations they purchase. Currently, the physical limitations of chip and integrated circuit technology are threats to the validity of this law. However, other

computing technologies are constantly being sought, such as biological and nano-technologies.

4   Interested readers may wish to consult the Japanese experience in harnessing knowledge assets in their corporations (Nonaka & Takeuchi, 1995). Examples from Japanese companies became a beacon to their American and European counterparts.

5   See, for example, Dertouzos (1997) who wrote: "First, humans deal with information on three levels. We receive it with all our senses. We process it with our nervous system and in a miraculous and largely unknown way with our brain. We also generate it as our brain commands our muscles to speak, scream, gesture, and type....Second, information can be a *noun* or a *verb*....Third, information is not the same as the physical thing that carries it" (p. 13). At this stage of knowledge management, there has not been, to my knowledge, a systematic effort to advance beyond establishing a typology of information and knowledge.

6   See some examples from the burgeoning business literature: Stewart, T. (2003). *The wealth of knowledge: Intellectual capital and the twenty-first century organization.* New York: Doubleday & Company; Hart, D. (Ed.). (2003). *The emergence of entrepreneurship policy: Governance, start-up, and the U.S. knowledge economy.* New York: Cambridge University Press; Drahos, P., & Braithwaite, J. (2003). *Information feudalism: Who owns the knowledge economy?* New York: The New Press; Stehr, N. (2002). *Knowledge and economic conduct: The social foundations of the modern economy.* Toronto: University of Toronto Press.

7   See, for example, Alai (2004), Fallis (2004), Audi (2002), Anderson et al. (2002), and Fuller (2001).

8   For example, Goldman (1999) and Geisler and Ritter (2003).

9   I use the definitions offered by Laudon and Laudon (2002).

10   Later in the book I refer to this entity as the "knower." The requirement for an external processor of the data is crucial to the definition, but leads to questions that philosophers had long explored—in particular, the issue of the existence of knowledge *outside* the knower. The arguments regarding knowledge can be extended in the taxonomy to the level of data. See, for example, Cross (2001).

11   The interested reader should consult: Shannon, C. (1948). A mathematical theory of communication. *Bell System Technical Journal, 27,* 379-423,

623-656; Weaver, W., & Shannon, C. (1949). *The mathematical theory of communication.* Urbana: University of Illinois Press.

[12] Ikujiro Nonaka and his colleagues defined knowledge in terms of actions because their frame of reference was the business corporation. Knowledge is viewed primarily as a tool in decision making and in executive action. At the level of the individual, this definition would refer to the person's *ability* (potential rather than actual) to act upon information received. Once such information acquires this utilitarian capability, it may be termed "knowledge." For additional sources, see Cover and Thomas (1991), Post and Anderson (2002), Goldman (1999), and Von Krogh, Ichizo, and Nonaka (2000).

[13] See, for example, Adams (2003), Cover and Thomas (1991), Meek (2003), and Parrini (1998).

[14] These are infinitely repeatable calculations. Turing argued that non-recursive functions cannot be computed by his machine, and this view was corroborated by Church (1941). The notion of "Turing-computable" suggests that, given time and ability of computers, any representation of the physical world can be mathematically computable. Also see Copeland and Proudfoot (1999).

[15] The reader who is interested in learning more about AI could consult the following sources: Kurzweil, R. (2000). *The age of spiritual machines: When computers exceed human intelligence.* New York: Penguin Books. Professor Kurzweil was a strong advocate for AI and a fervent believer in its unlimited potential. Also see: Stock, G. (2002). *Redesigning humans: Our inevitable genetic future.* Boston: Houghton Mifflin; Simon, H. (1996). *The sciences of the artificial* (3 ed.). Cambridge, MA: MIT Press; S., & Norvig, P. (2002). *Artificial intelligence: A modern approach.* Upper Saddle River, NJ: Prentice Hall; Hayles, K. (1999). *How we became posthuman: Virtual bodies in cybernetics, literature, and informatics.* Chicago: University of Chicago Press.

[16] AI has branched out to several areas, including machine learning, pattern recognition, game theory, parallel processing, genetic algorithms, and neural network analysis. For the Turing test, see: Turing, A. (1950). Computing, machinery, and intelligence. *MIND, 59*(3), 433-460. For additional reading and criticism of the future and capabilities of AI see: Simon, H. (1996). *The sciences of the artificial.* Cambridge, MA: MIT Press; Eberhart, R., Shi, Y., & Kennedy, J. (2001). *Swarm intelligence.* New York: Morgan Kaufmann; and Ned Block's critique of machine

functionalism in: Block, N. (1978). Troubles with functionalism. In C. Savage (Ed.), *Minnesota studies in philosophy of science, IX* (pp. 235-261). Minneapolis: University of Minnesota Press.

[17] See, for example, Luftman (2003), Post and Anderson (2002), Huysman and Wulf (2004), and Zuniga (2001).

[18] The interested reader may wish to consult, for example: Fallis (2004), Goldman (1999), Young (2004), and Nirenberg, Carbonell, Tomita, & Goodman (2000).

[19] In his Richard T. Ely lecture, Kenneth Arrow (1994) had argued that "knowledge and technical information have an irremovably social component, of increasing importance over time" (p. 8), and that "information may be supplied socially but to be used, it has to be absorbed individually. The limit on the ability to acquire information is a major barrier to diffusion" (p. 8).

[20] The interested reader should consult the following sources: Kahneman, D., Slovic, P., & Tversky, A. (1982). *Judgment under uncertainty: Heuristics and biases.* New York: Cambridge University Press; Kahneman, D., & Tversky, A. (Eds.). (2000). *Choices, values, and frames.* New York: Cambridge University Press; Festinger, L. (1957). *Theory of cognitive dissonance.* Stanford, CA: Stanford University Press. Other studies in the psychology of decision making and the utilization of information in human choices can be found in: Plous, S. (1993). *The psychology of judgment and decision-making.* New York: McGraw-Hill; Hirshleifer, J., & Riley, J. (1992). *The analytics of uncertainty and information.* New York: Cambridge University Press. Also: Raisinghani, M. (Ed.). (2004). *Business intelligence in the digital economy: Opportunities, limitations, and risks.* Hershey, PA: Idea Group.

Chapter II

Theories of Knowledge:
What We Know about What We Know

This chapter reviews the main streams of research on knowledge, assembled from a diversity of academic disciplines, such as philosophy (epistemology), philosophy of science, psychology, economics, and management and organization sciences. For the sake of continuity in the style in which this book is written, and for the comfort of the reader, I shall do my best to refrain from overusing esoteric terminologies of these streams or research movements. Terms such as "positivism," "phenomenology," "modernism," "reconstruction," and similar descriptions will be avoided, as will "post" and "neo" in conjunction with any of the above. Simply, the review will address what I believe to be the key arguments and the foundational components of prior scholarship and their contributions to the modeling of the structure of knowledge.[1]

It all began some two million years ago when one of our ancestors, named *Homo habilis,* found a way to make tools, thus paving the way for *Homo sapiens* (the "wise" human). As humans evolved, they continued to aggregate knowledge about their surroundings and increasingly engendered knowledge on how to survive in such environments, even how to master them. They developed a pool of knowledge on the making and use of tools and artifacts for hunting and gathering, and rudimentary language skills to maintain and diffuse these skills to their offspring.[2]

Some ten thousand years ago (it is not clear why), the Paleolithic age ended and the Neolithic age appeared.[3] Humans dramatically improved their tool-making, converting from hunting-gathering to food cultivation and animal husbandry as the key method for survival. To safeguard their growing populations and the surpluses of food and artifacts, the Neolithic humans developed communities, then cities, then the knowledge and technologies to administer them and to record their possessions.

Clearly, the knowledge our ancestors possessed was practical knowledge inherent in skills that allowed them to perform simple tasks that kept them alive and made them prosper. The questions posed by their inquisitive minds were perhaps directed toward the "how": How to start a fire? How to record the hunt on a cave's wall? How to prepare and use a hunting device?

With the growth of communities and the widespread appearance of farming, transportation, and storage of food surpluses, there was a need to deal with more complex stocks of knowledge. Solutions had to be found to problems never before encountered. The complexity of growing communities and the contemporaneous development of language and writing became evident forces that propelled people with public responsibilities to increasingly ask "why?" They observed patterns that were more complex than nature's rhythms of celestial movements, the seasons of the year, or the flooding calendars of mighty rivers.

To a large extent they procured the answers to "why" in divine or supernatural sources. But these solutions were not entirely satisfactory, thus forcing the ancients to engage in a more systematic pursuit of their natural surroundings. At first they transferred to their offspring their knowledge about hunting, gathering, tool making, and usage based on experience and imitation. Paleolithic humans gained knowledge from direct experience by their close proximity to nature. Suddenly, at some point some 15,000 years ago, there is a marked increase in applications and innovations in knowledge about farming and collective or social congregation of larger groups of humans sharing in the means of production and distribution of food and artifacts. Scholars who examined this change generally like the explanation of a spike or a revolutionary change in behavior caused by a cataclysmic event or some other unforeseen circumstance.[4] In this case explanations vary from dramatic changes to an explosion in population.

Practical knowledge, learned by hands-on experience and imitation, gave way to conceptual knowledge, transferred and learned by a method that is "one-step removed" from actual practice. This was done with the help of language, writing, and drawings. It meant that practical knowledge on how to start a fire or kill a prey with a sharp instrument was replaced by the notion or concept of creating fire with artifacts, and of obtaining food from prey by "long distance" hunting. Stories around the communal fire in the cave or other habitation may have been accompanied by movements that imitated an impressive hunting episode, but the listeners "were not there," thus had to *imagine* the encounter via notions of "hunting with instruments" and concepts of "bravery" and "communal good."

Perhaps the progress of knowledge, from practical to conceptual, was not so sudden. Even prehistoric humans had the capacity to imagine concepts such as deities, beyond the patterns they observed in nature. By cumulative practical knowledge they, and the Neolithic humans who followed, introduced small, yet meaningful improvements to the ways they procured food and shelter, and the means by which they governed the allocation of their resources.

Our human ancestors drew pictures from memory, requiring them to reconstruct shapes and events. They had to know and remember where food, in the form of edible vegetables and fruits, could be found and the dangers inherent in getting there and returning without harm. All this required accumulation of a variety of items of knowledge, and their positioning in a framework that would be amenable to reconstruction and to transfer to others.[5] It seems that early humans possessed the abilities to generate and process knowledge beyond the rudimentary practical or "hands-on" experience and imitation of key activities necessary for their survival.

Theories and Key Streams of Research on Human Knowledge

Gorman (2002) has offered an interesting summary of types of knowledge. He started with the main classification of knowledge as *explicit* and *tacit*.

The former is knowledge that can be clearly and perhaps even completely told or transferred by the knower. Tacit or implicit knowledge is that knowledge which is embedded in the knower, and which the knower is unable or unwilling to exchange.[6]

Gorman proposes four types of knowledge under these main categories. The first is "information," answering questions about *what*. This type of knowledge includes accretion of facts via memorization and reconstruction of reality from bits of information embedded in human memory. This type of knowledge is accomplished by using external memory aids, which help the knower to "find" the information needed.

A second type of knowledge refers to "skills," answering questions about *how*. This knowledge is also defined as procedural, so that algorithms may be established and the procedures codified. Gorman argues that such procedural knowledge under the explicit (declarative) category will be made of algorithms, whereas under the tacit (implicit) category, it consists of heuristics and hands-on knowledge.

The third type is "judgment," answering questions of *when*. The knower recognizes "that a problem is similar to one whose solution path is known and knowing when to apply a particular procedure" (p. 222). Under the explicit category, such knowledge relies on rules, whereas under the implicit (tacit) category, it is based on cases, mental models, and mental frameworks.[7]

The fourth type is "wisdom" knowledge, answering questions of *why*. Under the explicit category this knowledge relies on codes and under the tacit category it is based on moral imagination.

Gorman's summary is an excellent illustration of the focus on knowledge as a utilitarian tool or mechanism, employed to answer questions and to achieve individual and organizational goals. As models of knowledge emerge, stressing its functionality and utility, so have arguments that linked these models to the context of cultural and social influences.[8] Knowledge, it is argued, can be classified and its utility categorized only in the context of these cultural and social values and customs. Hence, perhaps we cannot, or should not, at this juncture, propose general models of how knowledge is utilized.

The focus on applications and utility that characterizes more recent scholarship on the nature of knowledge limits the exploration to somewhat vague, and certainly quite broad, taxonomies. As recent as this scholarship may be, it has not advanced our examination of the nature of knowledge to the level of its components. This is not only the gap in current scholarship, but has also been a constant gap in the models and theories offered by philosophers who studied epistemology. These philosophical works, beginning with Aristotle and flowering during the period of the "enlightenment," continue to this day. This body of work, although containing a variety of approaches, nonetheless illustrates a key stream in the study of knowledge with a focus on the overall nature of knowledge, its morphology, and its ethical implications. The basic elements of knowledge were not adequately explored. The emphasis was, and continues to be, on what constitutes knowledge and whether the knower indeed "knows." Thus, epistemologists are concerned with how we know and how we attest or recognize what we know. The other stream of scholarship, anchored in knowledge management, concerns the notion of knowledge and in particular its diffusion and manipulation.[9]

The Search for the Nature of Knowledge

Even this subtitle would require an entire book to adequately describe the research involved in such a search. I chose a selected set of authors, from Immanuel Kant to Karl Popper and F. Hayek in our times. In the two centuries I will attempt to cover, there has been a marvelous crop of scholars and philosophers who engaged in the study of knowledge. They represent a sundry array of viewpoints and approaches.

It would be unnecessary and certainly unwise to engage the reader in a full discourse of the various "schools of thought" and streams in philosophical and epistemological studies. I chose instead to briefly describe the main models, arguments, and findings of the key scholars.

The early philosophers concerned with knowledge had been the Greek scholars, known as Sophists. They posited that knowledge is wholly derived from experience. Plato (428-348 B.C.) disagreed with the Sophists. His theory of knowledge contended that knowledge based on sensorial experience is a low level of awareness. He argued that these are merely opinions and that the true level of knowledge or awareness is made of unchanging ideas or immutable forms that can be attained by reason and intellectual pursuit rather than by empirical experiences.[11]

Aristotle (384-322 B.C.) was a student of Plato, later also becoming a teacher in Plato's Academy in Athens. Aristotle initially supported Plato's view that abstract or ideal knowledge is indeed a superior mode of awareness, but he later contended that there are two ways of acquiring knowledge. The first applies to most forms of knowledge and can only be achieved from experience. Another mode is a deductive method, by which superior knowledge is gained by following the rules of logic, such as in the form of the syllogism.

These early Greek philosophers had framed the debate about the nature of knowledge for centuries to come. The core issue of the debate was the mode by which knowledge is acquired by the human mind. Is it by means of experience, through our senses, or by means of deductive reasoning? This debate, somewhat dormant in the Middle Ages (to an extent discussed by Thomas Aquinas, 1225-1274), was intensely reignited during the Age of Enlightenment in Europe, from 1600 to the early twentieth century.

Two groups of philosophers had emerged in the three centuries of this dichotomous intellectual struggle. One group, the rationalists, argued that human knowledge is obtained and can be verified by conducting a logical or rational exercise on principles of nature. These are given, self-evident postulates or, in their terminology, "axioms." From the Greek word for honor, axioms are accepted as true statements because of their intrinsic value, which one therefore must honor. The main method by which knowledge is gained, starting with these axioms, is the system of deduction.

The best known rationalists are Rene Descartes (1596-1650), Baruch Spinoza (1632-1677), and Gottfried Leibniz (1646-1716). The main assertion of this school of philosophy (concerning the issue of epistemology, or knowledge) was that the human mind has the ability to recognize and be cognizant of the world outside it by means of application of rational processes, without direct recourse to empirical experience.[12]

Descartes had argued that rational manipulations allow for the identification of universal principles or truths. These are indigenous to the mind and are independent of external events such as individual experiences. In the current terminology of knowledge management, Descartes proposed that we possess tacit knowledge of rational exercises that by deduction allow us to arrive at knowledge about all other aspects of the physical world. In other words, Descartes believed that we possess, in our mind, the formula that allows us by mental processes of rational thinking to know the physical world that exists outside of our own self.[13]

Descartes held that there is a fundamental separation between intellect and body, and that knowledge is based on absolutes derived from rational deduction. He therefore refused to accept any belief unless it was the product of rational examination, doubting even his own existence. Hence his famous

"Cogito, ergo sum" ("I think therefore I exist"). This rational deduction allowed him to declare it an axiom and to proceed from there to the deduction of the principles and laws of the natural world.[14]

Another rationalist was the Dutch philosopher Baruch Spinoza, the son of Portuguese Jews who emigrated to Holland. Spinoza lived a life of solitude and contemplation. Because of his philosophical ideas, he was excommunicated in 1656 by the rabbinical council of Amsterdam.

Spinoza argued that knowledge can be deduced from the basic laws and axioms, in the same way as in geometry and mathematics in general. He critically examined Descartes' duality of mind and matter, and arrived at the conclusion that mind and matter are two manifestations of the same phenomenon, existing in parallel trajectories. Hence, knowledge is itself a mode or a form of matter or substance. This conclusion did not solve the problem of the duality of mind and matter, but allowed Spinoza to suggest that by parallel existence, mind and matter "appear" to interact by our perception that they travel in the same wavelengths.[15]

Leibnitz also believed that a rational format or plan is responsible for the natural world. He proposed a system by which the physical world is made of "monads," which are items or centers of energy, acting as microcosmic representations of nature. They exist in harmony in light of the plan that God has pre-determined. Perfectly rational knowledge is to understand God's plan for the harmonious co-existence of the monads. However, the human mind is not capable of grasping such a perfectly divine plan and is therefore limited in its rational capability.[16]

On the other side of the quest for the nature of knowledge was a group of philosophers commonly known as *empiricists*. By chance rather than by cultural design, the most famous empiricists were British, whereas the key rationalists came from the continent. Empiricists believe that knowledge is primarily based on experience and our ability to sensorially capture the empirical world. As a school, the early empiricists rejected the notion of axioms and self-evident principles from which the mind can deduce most valued knowledge. John Locke (1632-1704) was an early proponent of this approach. He received his education at Oxford University and held public office, but never a professorial position. Locke's theory of knowledge is closely intertwined with his political theory. He was an ardent Protestant and strongly opposed the divine right of kings as being inconsistent with his philosophical belief that pre-existing ideas or notions are not valid. His theory of knowledge proposed that the human mind is in its origin a "tabula rasa" (an empty platform) into which sensorial inputs are imprinted as empirical manifestations of human experiences.[17] Since there are no pre-conceived ideas, Locke believed that each mind (hence each individual) is equal in his attempt to gain and to utilize knowledge. Although he died almost a century before the American Revolution, Locke's ideas featured prominently in the deliberations of the first Congress and in the drafts of the American Constitution.[18]

But the more influential empiricists were Berkeley and, particularly, Hume. George Berkeley (1685-1753) was a clergyman who taught at the Universities of Dublin and Oxford, where he is also interred. In his treatise on human knowledge, Berkeley rejected Locke's distinction between ideas and the physical world (empirical objects). He argued that knowledge is confined to ideas that we form in our mind about the empirical world. The physical world outside the human mind is irrelevant, since these things that are in such a world cannot be construed by the mind as concrete and real. The mind can only contemplate its own ideas.

Berkeley was deeply concerned with the skepticism and atheism of the philosophical approaches of his time. He therefore arrived at the conclusion that the thoughts in the human mind are there by transfer from a more able mind, that of God. He also argued that: "The ideas of sense are allowed to have more reality to them...but this is no argument that they exist without the mind" (Berkeley, 1957, p. 38).

Berkeley is considered the founder of "idealism" due to his belief that objects of the real world only exist if the human mind perceives them as such. What if the mind does not (temporarily) perceive the objects in the physical world outside the mind? Berkeley then argued that they are being perceived by God, hence at any given time objects are perceived. His famous phrase was: "*esse est percipi*" (to be is to be perceived). The knower's mind does not have evidence of the "true" existence of the physical world of objects outside the knower, because the knower perceives by means of a stream of sensorial inputs from this world. But these sensorial inputs are lodged in the mind—hence we are confined to the reality as it is perceived by and within the mind.

A more radical view of empiricism was advanced by David Hume (1711-1776). Born in Edinburgh, Hume spent many years in France, where he befriended Rousseau and other French scholars. Hume believed, as did Berkeley, that true knowledge of the natural world is impossible, and that knowledge is only possible by means of experience. The knower perceives such experiences, with all his flaws and subjectivity.[19] Hume's skepticism is generally exemplified by his view of causality and inductive reasoning. He doubted both. Laws of cause and effect, he argued, are mere beliefs, and there are no logical or rational grounds to draw any inferences from past events to the future.[20]

Hume made a very influential distinction between what he called "impressions" and "ideas," so that this distinction served as background to Kant's criticism of Hume and to the development of Kant's perspective of knowledge. Hume defined impressions as those experiences that we receive directly from our senses, as a sensorial representation of the external universe. Ideas, on the other hand, are those experiences that we know because we are able to extract them from impressions we had already experienced. These are, in a way, "derivatives" of the more powerful and real impressions, which are derived directly from our senses.[21]

Reconciliation and a Brilliant Step Forward

At the core of the dispute between the two schools searching for the nature of knowledge (rationalism and empiricism) was the distinction between the roles of sensorial inputs and rational manipulations of ideas, notions, and concepts. There was also a search for two distinct, yet complementary aims. One was "*What* is knowledge?" and the other was "*How* do we know?" The quest for what is knowledge followed a path of philosophical inquiry into "true" knowledge, and the human ability to "really" know the physical world outside the individual self and outside the mind.[22] This line of inquiry has produced perspectives on the ontology of knowledge, the ethical and religious implications of what it means to "really" know, and a fertile field of conjectures concerning the link that true knowing provides between mind and universe.

Very little came out of this line of inquiry that could illuminate the question of the structure of knowledge as an ontological entity (i.e., having its distinct form). The second school did not fare much better. The quest for understanding *how* we know followed a path of inquiry into how the mind processes whatever inputs it receives, from the external world (senses) and from itself (logic and reasoning).[23] As I described earlier, the two schools of thought (rationalists and empiricists) held extreme views, favoring either inputs from sensorial experiences *or* rational manipulations of ideas and concepts. There came a time when the need for the reconciliation and synthesis of these views became urgent and timely.

The first and monumental effort to reconcile and to synthesize the divergent approaches was the work of a professor at the German University of Konigsburg, Immanuel Kant (1724-1804). He wrote several books, two of which I will reference here: the *Critique of Pure Reason* (published in 1781) and the *Critique of Judgment* (published in 1790).

Kant was dissatisfied with the state of the philosophy of knowledge of his time. He believed that in order to reconcile between the distinct schools of thought, he needed to construct a unique and new framework of the nature of knowledge, with its very own terminology and concepts. Such a logical framework should also address the questions: What is knowledge? and How do we know?—that is, how the combined effects of sensorial inputs and rational manipulations are combined in the mind to create knowledge.[24]

Kant's framework is based on the distinction he makes (in the human mind) between perception and thinking. Perception deals with the sensorial inputs, and understanding deals with concepts. He classified concepts into three types: *a posteriori, a priori,* and *ideas.* Kant now faced the challenge of explaining how the two, seemingly diametrically opposed scenarios or models of the processing of knowledge indeed work in human cognition. This was not an easy task. He started by proposing that the human mind possesses "interactions," which are the criteria of *time* and *space* by which perceptions can be judged. Another attribute or capability of the human mind are a priori concepts called *categories.* So, the external world becomes knowable when sensorial perceptions are posited in the categories, within the criteria of the intuitions, thus forming judgments as to whether these sensorial inputs represent the external reality. The world outside the mind exists in the form of what Kant called "noumena," or the thing-in-themselves, but those are not knowable unless we can apply our perceptions of them for the categories.

There are, according to Kant, four groups of categories, each having three subcategories. These are:

- Quantity (unity, plurality, totality)
- Quality (reality, negation, limitation)
- Relation (substance and accident, causality and dependence, community or interaction)
- Modality (possible-impossible, existence-nonexistence, necessity-contingency)

By means of the categories, we are able to perceive objects in the physical world around us in a way that they seem to interact with each other and have causal relationships with each other and with us, the knower. This was the Kantian framework for applying empirical inputs in the creation of knowledge. But the mind also knows abstract notions or "ideas," which are higher-level constructs. Ideas, Kant posited, are not the outcome of sensorial or empirical perceptions that had been applied to the categories. Rather, they are the result of logical inference—the rationalist perspective.[25]

Kant also proposed two types of judgments: analytic and synthetic (a priori and a posteriori). Analytic judgments, propositions, or statements are inherently "true," hence they are known, but they do not provide us with knowledge about the world. Synthetic judgments are a "synthesis" between the knower and the world outside the knower. The statement: "The house on Main Street is a prairie-style architecture" is a synthetic judgment.

In Kant's framework of knowledge, synthetic a posteriori statements are based on the processing of sensorial data by the platform of categories. However, Kant struggled with the issue of *synthetic a priori* judgments or propositions. He argued that they do exist. The problem is that these judgments produce knowledge about the world without the input from sensorial data to the point where we have knowledge about this world that we are certain is true and known as we know analytic statements.

Kant argued that synthetic a priori knowledge is the mainstay knowledge in mathematics and in the sciences. General laws of science are not the result of sensorial inputs from our universe, but synthetic a priori knowledge that allows us then to organize our perceptions of the physical world into a meaningful set of connections.[26] As the individual knower applies the general rules to sensory perceptions, the Kantian categories allow the knower to identify the connections and form a meaningful (or knowable) and nonchaotic perception of the world. This, in essence, is what Kant called the "transcendental logic."

Kant's influence extended beyond his contribution to the scholarship on knowledge. He influenced the work of Marx, Hegel, Schopenhauer, Fries, Heidegger, Hayek, Popper, and a host of other philosophers and political scientists. His framework of how knowledge is processed in human cognition and his synthesis of the rationalistic and experiential perspectives turned out to be a very viable platform to understand the nature of knowledge, albeit also leading to selective criticism.[27]

From Kant to the Present

After Kant we find a hiatus in the pursuit of the nature of knowledge.[28] The emphasis has thus shifted from exploration of the *structure* of knowledge to a focus on the linguistic and symbolic *exchange* of knowledge. The comprehensiveness of Kant's scheme had a long-standing impact, so that scholars of knowledge were largely contented with examining the meaning, ramifications, and implications of Kant's contributions. Kant's scheme not only bridged the conflicting approaches to knowledge processing (positivism and empiricism), but also created a system that attempted to describe and explain the elements of human cognition. This was such an all-encompassing effort that it provided a platform for a diverse group of followers to pick and choose aspects of the scheme and build upon them their specific theories. Kant's logical framework was like a "supermarket" of possible avenues for exploration: political or philosophical, ethical or economic, religious or so-ciological. The spin-offs from Kant's framework were essentially limitless, thus occupying the attention of scholars for decades afterwards.

In the past two centuries since Kant, the exploration of the nature of knowl-edge was carried out by a mixed bag of sociologists (such as Durkheim), social-anthropologists (Levi-Strauss), communication scientists, psycholo-gists, and more recently, information scientists. This trend inevitably led to the emergence of the linguistic philosophy of knowledge and to Wittgenstein, Quine, Russell, and Chomsky.

Analytic and Linguistic Approaches to Knowledge

The general trend of the pursuit of the structure of knowledge in the twentieth century had focused on propositions or statements and their characterization of knowledge. The emphasis was on how people *exchange* and *communicate* what they know, rather than the structure of what they know. Statements in the language that people use include concepts and notions in their entirety, therefore they do not require a more in-depth exploration into what makes these statements bearers of knowledge. This means that the onus is now on determining whether such knowledge-laden statements are true or false, and the modes or procedures that one would use to ascertain their veracity.[29]

Ludwig Wittgenstein (1889-1951) was a student of Bertrand Russell (1872-1970). Both may be credited with founding the school in philosophy known as *logical positivism.* Russell, an ardent mathematician, believed that the complex physical world can be explained by simplifying its components into precise and meaningful propositions. He named them *atomic proposi-tions.*[30] In cooperation with Alfred North Whitehead (1861-1947),[31] Russell introduced mathematical symbols to simple propositions that describe the physical world. He argued that such logical propositions are meaningful, in that they correspond to the elements of nature, in what Russell called *logical atomism.* This one-to-one correspondence between the logic of language and the universe allows us to gain knowledge about our universe and to character-ize it in a form that is meaningful and exchangeable with others.

Wittgenstein was strongly influenced by Russell. He and other philosophers of his time (such as Mach and Schlick) formed what was known as the Vienna circle or school of linguistic philosophy. Wittgenstein believed, as did Rus-sell, that language can be reduced to elementary propositions that describe the physical world. These propositions are meaningful when they describe *facts,* such as propositions of scientific knowledge.

However, in a later book, *Philosophical Investigations,* Wittgenstein (1968) recognized the different uses of language to give meaning to scientific analysis, as well as in religious, commercial, and other uses. He argued, therefore, that propositions must be understood within the context in which they are utilized. He introduced the notion of "language games" that people play by using language as a tool in their dealings with their universe (Mounce, 1990).

More recently, Willard Quine (1908-2000) extended Wittgenstein's notions of the uses of language (Quine, 1951). He criticized the distinction made between synthetic and analytic statements or propositions. Quine addressed the issue of how one knows the world by suggesting that the use of language and the choices one makes in linguistic varieties have a great effect on the way one perceives the external world. Knowledge is therefore a reflection of the use of language.

The linguistic perspective in the pursuit of the structure of human knowledge has championed this search to the extent that it became bogged down with issues of the form and usage (functionality) of language. Even the study of cognition has a strong bias toward the role of language in the processes of the human mind.[32]

Noam Chomsky is a leader in the study of linguistics. He challenged existing theories on the structure of language by suggesting that such theories should also explain how language is used in processes of the human mind. His con-tribution had to do with the distinction between the knowledge of language skills and the specific uses that humans make of these skills. He spoke of "generative grammar," which is the link between the structure of language and its applications in human cognition (Chomsky, 1972, 2002).

These theories that emphasized the linguistic perspective have side-tracked the pursuit of the nature and structure of knowledge. The focus was on propositions, statements, and the nature and verifiability of complex descriptions of nature. At the level of language, these scholars already started from a complex point of concepts and notions that can be expressed in words and arranged in propositions and statements. Moreover, this line of scholarship had led them to believe that cognitive processing of knowledge is in the form of propositions.[33]

At issue was the seemingly conflicting view of how the mind perceives the external world. Does the human mind form a pictorial or *analog* image of external reality, or does it perceive it by means of statements that indirectly describe the external world (*digital* representation)?[34] Although this conflict continues to exist, some recent studies have attempted to better explore this issue by focusing on human problem solving and its similarity with how computers operate.

Management, Problem Solving, and Psychology

Comparison of human cognition with the newly developed computers has been a catalyst to a large portion of scholarship in the areas of decision making, problem solving, and psychology. Some early work after the Second World War was carried out by Herbert Simon (1916-2001) and his colleagues at Carnegie Mellon University.[35] They contributed to the generation of the areas of artificial intelligence, automata, and robotics. They endeavored to create computer programs and machines or instruments that are capable of reasoning that approaches human thought.

Simon also developed the concept of *bounded rationality.* He argued that in human (particularly in managerial) decision-making processes, it is impossible to gather, absorb, and analyze all the information one would need to make a completely rational decision so that it would maximize the benefits from such a decision (Simon, 1991). Instead, Simon suggested that managers make decisions on the basis of the amount and quality of information that satisfies their level of comfort with the decision and its outcomes—rather than continually pursuing the "maximized" or "optimized" level of decision making.

More recently, there have been developments in psychology and managerial cognition to address issues of human perception, cognition, and imaging. Kahneman, Slovic, and Tversky (1982), for example, demonstrated that humans make choices and exhibit preferences in uncertain environments based on mental or psychological representations that are generally different from logical rules of inference of a rational decision theory.[36]

The State of Affairs

The study of human knowledge and its ramifications into larger systems such as organizational and managerial systems of knowledge has been a multi-disciplinary effort. There has been little cross-fertilization or inter-disciplinary research (Hazlett, McAdam, & Gallagher, 2005; Nonaka & Teece, 2001; Patriotta, 2004).

This has constrained researchers to delve ever more deeply into their own conceptual frameworks and their parochial methodologies. The emphasis has largely been on *how* knowledge is processed, rather than an exploration of its elemental structure and its unit of analysis. The adherence to the chains of data–information–knowledge has also led to a focus on the information–knowledge flow. As information scientists extended their exploration into the concept of knowledge (as a natural continuation of the chain), they also inflicted upon the study of knowledge the ideas, methods, and focus of information science.

The combination of a diversified disciplinary landscape and the emphasis on process and later on relevance and applications has created a state of affairs in which knowledge has become an orphaned creature of the massive research effort in the fields of information, cognitive sciences, and management. Even in the emerging literature that specifically targets knowledge management, the focus of research remained within processes, value, and utilization (Agarwal & Lucas, 2005; Cheng et al., 2004; Lockett & McWilliams, 2005).

This book is one small step aimed at a remedy for this state of affairs. The first two parts of the book focus on the basic unit of structure of knowledge and on a model of its progress. The latter part of the book follows the extant literature by linking the model thus developed to the world of knowledge systems and their applications. If I defiantly stray from the main in the initial half of the book, I then obsequiously return the narrative to the mainstream body of research of applications and utilization.

What do We Know?

This intense intellectual effort we have witnessed in the recent past has not resulted in much progress in the quest for understanding the structure of knowledge.[37] The combination of research on linguistics and semiotics, and on rationality and the architecture of reason has been bogged down in trends that lead away from investigation of knowledge, its structure, and its dynamics.

A very revealing book by Zeno Vendler (1972) portrays a good illustration of the state of affairs in our understanding of knowledge. Following Chomsky, Vendler attempted to relate language to ideas or mental images. He concluded that:

"One could argue that although this theory might explain the ease children display in learning a language and thus may have some importance for scientific psychology, with respect to the philosophical problem of ideas it offers no solution—it merely pushes the problem further back in time. By suggesting that these ideas are native in individual humans (as we know them now), one does not say anything about the absolute origin of these ideas...In consequence, we are still up in the air concerning their relation to the world." (p. 217)

Vendler argues that such native ideas are subject to human evolution, and are a tool with which human beings are able to confront and understand their external reality. He quipped: "It is bad enough that we are born as a 'naked ape' in the body; why should we start out with a *tabula rasa* for a mind as well?"

Vendler is correct. Although progress has been made in several ancillary intellectual areas, we have "pushed the problem back in time." As we had embarked in recent years on the study of propositions and their linguistic and rational meanings, we are still much in the dark on what constitutes knowledge, how it is structured, and what its elemental constituents are.

We do recognize that the structure of human knowledge is composed of two major elements: the processing of signals from our environment and the conceptual tools (ideas, categories, etc.) with which we undertake such processing. We also recognize the roles that beliefs, biases, perception, and other psychological phenomena of our mind play in processing inputs from the external world. Finally, we understand the role of language, semantics, and semiotics in portraying and describing the external world and our knowledge of what we consider to be reality.

Emerging Interest in the Working of the Mind

Since the mid-1990s there has been a surge in the levels of both popular and academic interest in the human mind. This resulted in a flurry of books and scholarly publications.[38] This phenomenon may be credited to the converging effects of three factors. The first was the increasing ubiquitousness of medical imaging and diagnostic technologies. There has been a dramatic leap in the uses of such technologies as computed tomography (CT), magnetic resonance imaging (MRI), positron emission tomography (PET), single photon emission computed tomography (SPECT), and X-ray tomography. These technologies offered much more advanced, more focused, and more discriminating pictures of the brain, its activity levels, and the positioning of selected emotions and cognitive functions within the geography of the brain.

The second factor has been the innovative developments in research and applications of the cognitive sciences.[39] Increasingly, there were discoveries of various aspects of cognitive impairment, such as Alzheimer's. These advances have captivated the public's imagination and have diverted the limelight to the functioning and mysteries of the human brain.

Thirdly, the unparalleled developments in human genetics have contributed to the overall revolutionary belief in the public opinion that humanity—driven by scientific progress—is on the verge of finding cures for many hitherto less understood and untreatable maladies. This belief had been extended to the complexities of the human brain and to its deficiencies and pathologies. Such a phenomenon gained prominence in particular as the "baby boomers" began to age.[40]

The combined impacts of these factors have led to clinical advances in the imaging of the brain and the resulting improvements in diagnostic techniques and successes in the discriminate identification of cognitive impairment.[41] In addition, advances in research into the cognitive sciences and new discoveries in pharmacology have created a host of "miracle" drugs for the treatment of ailments such as depression, eating disorders, and schizophrenia.

As scientists continue their explorations into the workings of the human brain, we are entering in the early years of the 21st century into a clinical revolution of discoveries in diagnostics and therapeutics.

In parallel, there have been advances in economics, management, and organization theory further discussed in this book. These disciplinary areas identified the emergence of the knowledge economy and knowledge workers as the new assets of the post-industrial world. Within the span of a few years, there has been a rapid growth in the interest by academics and practitioners of how to harness knowledge and how to construct effective knowledge systems for use by managers and their work organizations. The combination of these phenomena is sorely wanting to deal with our basic understanding of knowledge.

The complex structure and ubiquitousness of knowledge systems are some of the key forces that challenge us to "look inside the box" and to gain a better understanding of how knowledge is structured. To this end I embarked on the journey described in this book. The starting point is the next chapter, where I examine the seeds of knowledge: What is the basic unit of that which we call "knowledge?"

References

Agarwal, A., & Lucas, H. (2005). The information systems identity crisis: Focusing on high-visibility and high-impact research. *MIS Quarterly, 29*(3), 381-398.

Allison, H. (2004). *Kant's transcendental idealism: An interpretation and defense.* New Haven, CT: Yale University Press.

Anderson, J., & Lebiere, C. (1998). *The atomic components of thought.* Mahwah, NJ: Lawrence Erlbaum.

Audi, R. (2002). *The architecture of reason: The structure and substance of rationality.* New York: Oxford University Press.

Beck, L. (1996). *A commentary on Kant's critique of practical reason.* Chicago: University of Chicago Press.

Berkeley, G. (1957). *A treatise concerning the principles of human knowledge.* New York: MacMillan.

Brook, A. (2001). *Knowledge and mind: A philosophical introduction.* Cambridge, MA: MIT Press.

Bushkovitch, A. (1974). Models, theories, and Kant. *Philosophy of Science, 41*(3), 86-88.

Carpenter, H. (2004). *The genie within: Your subconscious mind, how it works and how to use it.* San Diego: Anaphase II.

Cassirer, E. (1950). *Problem of knowledge: Philosophy, science, and history since Hegel.* New Haven, CT: Yale University Press.

Cheng, P. et al. (2004). Knowledge repositories in knowledge cities: Institutions, conventions, and knowledge subnetworks. *Journal of Knowledge Management, 8*(5), 96-112.

Chomsky, N. (1972). *Language and mind.* Stamford, CT: Thomson.

Chomsky, N. (2002). *Syntactic structures* (2nd ed.). New York: Walter de Gruyter.

Clark, A. (2001). *Mindware: An introduction to the philosophy of cognitive science.* New York: Oxford University Press.

Coff, R. (2003). The emergent knowledge-based theory of competitive advantage: An evolutionary approach to integrating economics and management. *Managerial and Decision Economics, 24*(4), 245-264.

Cooper, D., Mohanty, J., & Sosa, E. (Eds.). (1999). *Epistemology: The classic readings.* London: Blackwell.

Dancy, J., & Sosa, E. (1994). *A companion to epistemology.* Malden, MA: Blackwell.

Demopoulos, W. (2003). On the rational reconstruction of our theoretical knowledge. *British Journal for the Philosophy of Science, 54*(3), 371-389.

Doran, R. (1994). *Lonergan and Kant: Five essays on human knowledge.* Toronto: University of Toronto Press.

Dowling, J. (2000). *Creating mind: How the brain works.* New York: W.W. Norton & Company.

Fetzer, J. (Ed.). (1991). *Epistemology and cognition.* Boston: Kluwer Academic.

Fisch, M. (1994). Toward a rational theory of progress. *Synthese, 99*(2), 277-304.

Foss, N. (2003). Bounded rationality and tacit knowledge in the organizational capabilities approach. *Industrial and Corporate Change, 12*(2), 185-197.

Glisby, M., & Holden, N. (2003). Contextual constraints in knowledge management theory: The cultural embeddedness of Nonaka's knowledge-creating company. *Knowledge and Process Management, 10*(1), 28-36.

Goldman, A. (1986). *Epistemology and cognition.* Cambridge, MA: Harvard University Press.

Goldman, A. (1988). *Epistemology and cognition.* Cambridge, MA: Harvard University Press.

Gorman, M. (2002). Types of knowledge and their roles in technology transfer. *Journal of Technology Transfer, 27*(3), 219-231.

Gurian, M. (2004). *What could he be thinking?: How a man's mind really works.* New York: St. Martin's Griffin.

Gustavson, B. (2001). Towards a transcendent epistemology of organizations. *Journal of Organizational Change Management, 14*(4), 352-378.

Guyer, P. (1987). *Kant and the claims of knowledge.* New York: Cambridge University Press.

Haack, S. (1996). Precis of evidence and inquiry: Towards reconstruction in epistemology. *Philosophy and Phenomenological Research, 56*(3), 611-614.

Harman, G. (2002). Reflections on knowledge and its limits. *Philosophical Review, 111*(3), 417-428.

Hars, A. (2001). Designing scientific knowledge infrastructures: The contribution of epistemology. *Information Systems Frontiers, 3*(1), 63-73.

Hazlett, S., McAdam, R., & Gallagher, S. (2005). Theory building in knowledge management: In search of paradigms. *Journal of Management Inquiry, 14*(1), 31-43.

Heyes, C., & Hull, D. (2001). *Selection theory and social construction.* Albany: State University of New York Press.

Hodge, R., & Kress, G. (1988). *Social semiotics.* Ithaca, NY: Cornell University Press.

Holmes, E. (2002). *How to use the science of mind.* Burbank, CA: Science of Mind.

Hume, D. (2000). *A treatise of human nature.* New York: Oxford University Press.

Hutton, J. (1999). *Investigation of the principles of knowledge.* Bristol, UK: Thoemmes Press.

Jacobs, J. (1999). *Depravity of knowledge: The Protestant reformation and the disengagement of knowledge from virtue in modern philosophy.* Aldershot, UK: Ashgate.

Kahneman, D., Slovic, P., & Tversky, A. (Eds.). (1982). *Judgment under uncertainty: Heuristics and biases.* Cambridge, UK: Cambridge University Press.

Kant, E. (1999). *Critique of pure reason.* Cambridge, UK: Cambridge University Press.

Kant, I. (1970). *Critique of judgment.* New York: The Free Press.

Kemp-Smith, N. (2003). *A commentary to Kant's critique of pure reason* (2nd ed.). New York: Palgrave MacMillan.

Kincaid, H. (1996). *Philosophical foundations of the social sciences.* New York: Cambridge University Press.

Kotulak, R. (1997). *Inside the brain: Revolutionary discoveries of how the mind works.* Riverside, NJ: Andrews McMeel.

Larson, C., & McLauchlin, L. (2003). *How the mind words.* Bloomington, IN: Authorhouse.

Liang, T. (1992). A composite approach to automated induction of knowledge for expert systems design. *Management Science, 38*(1), 1-17.

Lockett, A., & McWilliams, A. (2005). The balance of trade between disciplines: Do we effectively manage knowledge? *Journal of Management Inquiry, 14*(2), 139-150.

Margolis, H. (1993). *Paradigms and barriers: How habits of mind govern scientific beliefs.* Chicago: University of Chicago Press.

McDermid, D. (2002). Schopenhauer as epistemologist: A Kantian against Kant. *International Philosophical Quarterly, 42*(2), 209-229.

McLelland, J., & Dorn, H. (1999). *Science and technology in world history.* Baltimore: Johns Hopkins University Press.

Meek, E.L. (2003). *Longing to know: The philosophy of knowledge for ordinary people.* Wheaton, IL: Brazos Press.

Milmed, B. (1961). *Kant and current philosophical issues & some modern developments of his theory of knowledge.* New York: New York University Press.

Moinar, G. (1999). Are dispositions reducible? *Philosophical Quarterly, 49*(194), 1-17.

Mokyr, J. (2002). *The gifts of Athena: Historical origins of the knowledge economy.* Princeton, NJ: Princeton University Press.

Mounce, H. (1990). *Wittgenstein & Tractatus: An introduction.* Chicago: University of Chicago Press.

Nagel, J. (2000). The empiricist conception of experience. *Philosophy, 75*(293), 345-376.

Newell, A. (1990). *Unified theories of cognition.* Cambridge, MA: Harvard University Press.

Newell, A., & Simon, H. (1972). *Human problem solving.* Englewood Cliffs, NJ: Prentice Hall.

Nickles, T. (Ed.). (1980). *Scientific discovery, logic, and rationality.* Boston: Kluwer Academic.

Nonaka, I., & Nishiguchi, T. (2001). *Knowledge emergence: Social, technical, and evolutionary dimensions of knowledge creation.* New York: Oxford University Press.

Nonaka, I., & Teece, D.J. (2001). *Managing industrial knowledge: Creation, transfer and utilization.* Thousand Oaks, CA: Sage.

Norman, D. (1993). *Things that make us smart: Defending human attributes in the age of the machine.* Reading, MA: Addison-Wesley.

Owens, J. (1992). *Cognition: An epistemological inquiry.* West Lafayette, IN: University of Notre Dame Press.

Patriotta, G. (2004). On studying organizational knowledge. *Knowledge Management Research and Practice, 2*(1), 3-12.

Piaget, J. (1972). *Psychology and epistemology: Towards a theory of knowledge.* New York: Penguin Press.

Polanyi, M. (1974). *Personal knowledge: Towards a post-critical philosophy.* Chicago: University of Chicago Press.

Quine, W. (1951). *Mathematical logic* (revised ed.). Cambridge, MA: Harvard University Press.

Rehder, B., & Hastie, R. (2001). Causal knowledge and categories: The effects of causal beliefs on categorization, induction, and similarity. *Journal of Experimental Psychology, 130*(3), 323-360.

Rescher, N. (1990). *Baffling phenomena and other studies in the philosophy of knowledge and valuation.* London: Rowman & Littlefield.

Robertson, L., & Sagiv, N. (2004). *The cognitive neuroscience of synaesthesia.* New York: Oxford University Press.

Russell, B. (1929). *Our knowledge of the external world.* Chicago: Open Court.

Russell, B. (1994). *Human knowledge: Its scope and limits.* London: Routledge.

Russell, B. (1996). *The principles of mathematics* (reissue ed.). New York: W.W. Norton & Company.

Simon, H. (1991). Bounded rationality and organizational learning. *Organization Science, 2*(2), 125-134.

Simon, H. (1996). *The sciences of the artificial* (3rd ed.). Cambridge, MA: MIT Press.

Small, G. et al. (2006). PET of brain amaloid and tau in mild cognitive impairment. *New England Journal of Medicine, 355*(25), 2652-2663.

Tallman, S., Jenkins, M., Henry, N., & Pinch, S. (2004). Knowledge clusters and competitive advantage. *Academy of Management Review, 29*(2), 258-271.

Thompson, L., Levine, J., & Messick, D. (Eds.). (1999). *Shared cognition in organizations: The management of knowledge.* Mahwah, NJ: Lawrence Erlbaum.

Vendler, Z. (1972). *Res cogitans: An essay in rational psychology.* Ithaca, NY: Cornell University Press.

Whitehead, A. (1979). *Process and reality* (2nd ed.). New York: The Free Press.

Whitehead, A. (1985a). *Modes of thought* (reissue ed.). New York: The Free Press.

Whitehead, A. (1985b). *Symbolism: Its meaning and effect.* New York: Fordham University Press.

Wilson, E. (1998). *Consilience: The unity of knowledge.* New York: A. Knopf.

Wittgenstein, L. (1968). *Philosophical investigations.* Oxford, UK: Blackwell.

Zuniga, G. (2001). Ontology: Its transformation from philosophy to information systems. *Proceedings of the International Conference on Formal Ontology in Information Systems,* Ogunquit, ME.

Endnotes

¹   Readers may consult summaries of the current state of the art in, for example, Meck (2003), Mokyr (2002), Liang (1992), Hars (2001), and Nickles (1980).

²   This group of prehistoric humans are called "Paleolithic" meaning ancient (Paleo) and stone (lithas). As they evolved and improved their knowledge, they transformed themselves from hunters-gatherers to producers of food and animal husbandry, thus leading to the acquisition of surpluses, acute growth in population, thus to larger communities and public administration. See, for example, McLelland and Dorn (1999).

³   From the Greek: New (neo) and stone (lithos).

⁴   See, for example, Tattersall, I. (1993). *The human odyssey: Four million years of human evolution.* Upper River Saddle, NJ: Prentice Hall; Burenhult, G., & Thomas, D. (Eds.). (1993). *The first humans: Human origins and history to 10,000 B.C..* New York: Harper-Collins. Also see the classic book: Campbell, B., & Loy, J. (1999). *Humankind emerging* (8th ed.). Upper Saddle River, NJ: Pearson, Allyn & Bacon. Also see: Price, D., & Feinman, G. (2000). *Images of the past.* New York: McGraw-Hill.

⁵   Norman (1993), for example, refers to this ability as *accretion* of facts. His term is similar to my notion of *cumulation* of knowledge.

⁶   For further interest, the reader should consult Polanyi (1974) and Nonaka and Nishiguchi (2001). Polanyi defined tacit knowledge as "personal knowledge" which perhaps cannot be totally transferred from the knower to others.

7   For additional and more detailed reading, see Foss (2003), Coff (2003), and Hodge and Kress (1988).

8   For example, Glisby and Holden (2003), Tallman, Jenkins, Henry, and Pinch (2004), and Gorman (2002).

9   I have little doubt that philosophers will not agree with these statements. The purity of their search cannot be jeopardized with the quest for utility of the knowledge management folks. An important part of epistemological studies is guided by rational and logical controls. Yet, however elegant and self-contained the methodology and however noble the aims, in my view epistemologists swayed very little from the key questions of philosophy's main quest. The study of knowledge was more in terms of an illustration of broader philosophical issues, and the models thus developed served as vehicles to demonstrate larger issues. See, for example, Haack (1996) and Nagel (2000).

10   This approach may draw some criticism from scholars in the disciplines of philosophy and epistemology. Such a summary may be perceived as a "cheapening" of a serious area of scholarship. I reiterate, however, that this book is targeted at a wide audience and that this chapter serves only as a general introduction, rather than a treatise on epistemology.

11   Readers may be interested in his *Republic,* in which Plato offers the famous myth of the cave.

12   For initial review of this school, see, for example, Brook (2001), Fisch (1994), Owens (1992), Fetzer (1991), Goldman (1988), Haack (1996), and Dancy and Sosa (1994).

13   This interpretation of Descartes' work is my personal view and may not be in total agreement with scholars who specialize in rationalism or Descartes. Throughout this book I have invoked my own understanding of the various streams in philosophy and epistemology, to a degree of interpretation that would allow the reader to have a broad grasp of the work of others—prior to the description of my own model of how knowledge is structured and how it progresses. For additional readings, see Audi (2000), Cooper, Mohanty, and Sosa (1999), and Kincaid (1996).

14   Descartes also arrived at the existence of God and ethical laws that are the product of rational manipulations.

15   This also led Spinoza to believe that mind, intellect, matter, and therefore the physical world and God are all manifestations of the same

phenomenon. Hence, individuals are rationally bound to cooperate with each other (ethics) and all, people and nature, are part of a holistic view in which God's relations with the world and with human beings is an intellectual or rationalistic tie. For this, among other beliefs, he was excommunicated.

[16]   As I describe later in the book, this view is similar to the work by Herbert Simon and the Carnegie Mellon University group. They proposed the concept of "bounded rationality," which suggests that humans are unable to search, absorb, and analyze all the information they would need in order to make perfectly rational decisions that maximize or even optimize their decision objectives. They will resort to "satisficing," that is, decisions that are not the best but are "good enough" or "satisfactory" under the existing constraints. This view is similar to the way Leibniz considered the limits of human understanding—although, of course, arriving at this view from a totally different perspective.

[17]   See his most influential book: Locke, J. (1994). *An essay concerning human understanding*. Amherst, NY: Prometheus Books. Locke published this book in 1690.

[18]   Locke did not specify how sensorial inputs are accumulated in the mind nor how they progress and develop. This task was left to Immanuel Kant.

[19]   See Hume (2000) and Hume, D. (1999). *An enquiry concerning human understanding*. New York: Oxford University Press. Also see Nagel (2000).

[20]   Hume was very emphatic in his writings. He suggested that rational claims, not founded on experience or that are not directly perceived by the knower, must be destroyed. He further suggested that the individual knower has not a true knowledge of himself, as he is simply a depository of many different perceptions of reality.

[21]   Hume's ideas and his strong doubts that causality is even possible have influenced the use of heuristics and statistical analysis in contemporary scientific methodology. Hume essentially demolished the belief of the scientific revolution that empirical research and inductive reasoning can yield knowledge about basic principles of the universe.

[22]   The philosophy literature has been very prolific in discussing this age-old topic of whether we "truly" know, and what is the "true" knowledge and its essence. For a glance at this literature, see, for example: Moinar

(1999), Haack (1996), Zuniga (2001), Hutton (1999), Jacobs (1999), and Demopoulos (2003).

[23] In recent years the focus and terminology have shifted to "cognition" and its role in processing information and creating knowledge. See, for example: Anderson (1983), Clark (2001), Thompson, Levine, and Messick (1999), Foss (2003), and Dancy and Sosa (1994).

[24] The literature dealing with the work and the persona of Kant is very prolific and diverse. The reader will get a flavor by examining the following examples: Allison (2004), Guyer (1987), Kemp-Smith (2003), Beck (1996), and Nagel (2000).

[25] I received Kant's *Critique of Pure Reason* as a gift when I was thirteen. I have been exploring Kant's framework ever since. To some degree my model rests on Kant's framework. Those who will praise the model I advance in this book may point with satisfaction to the Kantian continuity. Detractors of my model may define it as mere extension of Kant's work. Both may be correct. So be it!

[26] Kant's term for this phenomenon of cognition was: "synthetic unity of the sensory manifold." In essence, this is the effort of clustering to which I refer in Chapters IV and V. For additional reading, see: Allison (2004), Guyer (1987), Beck (1996), Harman (2002), Rehder and Hastie (2001), Gustavson (2001), and McDermid (2002).

[27] See, for example, Cassirer (1950) and McDermid (2002). Also see Kemp-Smith (2003), Beck (1996), Piaget (1972), Owens (1992), Milmed (1961), Bushkovitch (1974), and Doran (1994).

[28] Until now this book continues in the search for the structure of knowledge and the components of its processing in cognition.

[29] For additional readings, see, for example: Heyes and Hull (2001), Hutton (1999), Margolis (1993), Rescher (1990), and Wilson (1998).

[30] See Russell (1929, 1994, 1996).

[31] See Whitehead (1979) and *Modes of Thought* (1985) and *Process and Reality* (1985).

[32] See, for example, two recent representative articles in the journal *Cognition*: Musolino, J. (2004). The semantics and acquisition of number words: Integrating linguistic and developmental perspectives. *93*(1), 1-41; Wang, M., Koda, K., & Perfetti, C. (2004). Language and writing are both important in learning to read: A reply to Yamada. *93*(2), 133-137.

[33]   See, for example, Anderson and Lebiere (1998), Newell (1990), and Simon (1996).

[34]   For a good discussion of these issues, see Goldman (1986).

[35]   See, for example, Simon (1996), Newell and Simon (1972), and Newell (1990).

[36]   Their conclusions reinforce the existing divergence between rationality and empiricism, with an added component of heuristics and bias that seem to be embedded in human cognition.

[37]   The effort in the area of management cognition was particularly disappointing. Much of the outcome focused on perception and little else related to the structure or processing of knowledge by managers. My own doctoral dissertation examines the manner in which managers are able to cluster empirical inputs into conceptual constructs. Although this work was done over a decade before the emergence of managerial cognition as a distinct research area in the managerial sciences, it still remains somewhat secluded from the mainstream of this research effort.

[38]   See, for example: Carpenter (2004), Dowling (2000), Gurian (2004), Holmes (2002), Kotulak (1997), and Larson and McLauchlin (2003). The majority of these books describe the emerging innovations in diagnostics and imaging of the human brain.

[39]   See, for example, Robertson and Sagiv (2004).

[40]   In the paraphrased words of the former Surgeon General of the United States, Dr. Charles Everett Koop, this generation of those born in the decade following the Second World War do not consider death an option. They require the best that the healthcare delivery system can offer, and strongly endorse the blessings of science and technology as providers of clinical miracles in diagnostics and therapeutics of all diseases of the body and the mind.

[41]   Small et al. (2006), for example, had used positron-emission tomography and magnetic resonance imaging on 83 volunteers who reported memory problems. The researchers were able to diagnose—with these images—patients with Alzheimer's disease and patients with only mild cognitive impairment.

Chapter III

Seeds of Knowledge:
Nuggets, Memes, and the Search for the Basic Unit

What is the basic unit of knowledge? To answer this pesky query means to also reveal what *is* knowledge and perhaps even what is the *structure* of knowledge. In such a pursuit we should start with some definitions of types and forms of knowledge, so that we can possibly gain desired common ground. In the previous chapter I discussed the recent focus on propositions and language as descriptors of knowledge. These are active at the level of words, concepts, and even complex notions, such as "belief" and "justification."

There have been several attempts to distinguish between "knowledge" and "to know," as well as distinct definitions of knowledge as "warranted belief" or "actionable information" (e.g., Werkmeister, 1948; Harman, 2002; Dewey, 1977; Cadamer, 1977). Table 1 shows the intersect between the *descriptors* of knowledge and their *attributes.*

Table 1 shows, for example, that knowledge sometimes has been described as "functional," so that it serves some human purpose or is a tool in human

Table 1. Descriptors and attributes of selected perspectives of knowledge

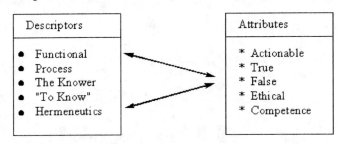

activities. Such knowledge-as-function may be "true" (when considered through the lens of propositions) or "false," or serve an ethical function, or be viewed as human competence.

The distinction between the knower and to know has been extended from simply being an issue of terminology (Williamson, 1999) to an examination of cognitive processes and the internal representation of human knowledge. Hence this extension attempts to link the knower and the process of what constitutes to know (Harman, 2002).

In his book on creative evolution, published in 1911, Henri Bergson (1859-1941) succinctly summarized the relationship between the knower and knowledge, or in his terms, between intelligence and reality (see Bergson, 1998). He argued that: "Human intelligence, as we represent it, is not at all what Plato taught in the allegory of the cave…To act and to know that we are acting, to come into touch with reality and even to live it, but only in the measure in which it concerns the work that is being accomplished and the furrow that is being plowed, such is the function of human intelligence" (p. 191).

I cite Bergson because he is an excellent example of a perspective of knowledge that views it as a function of human action, exercised at the level of human cognition, intelligence, or the sphere of conceptual thinking. Kant's distinction between the "things-in-themselves" (the "true" reality of which we do not know) and our perception of reality as "knowable" reality is maintained by those scholars who view knowledge either as a function or at the conceptual level of analysis. This has led to implications in political theory and other social sciences and to a field of "history of ideas" in which ideas or complex concepts are traced through history, geography, and social environments to determine their diffusion, growth, absorption, and transformation.[1] These ideas or concepts are also values embedded in people and their societies.

There is therefore a direct link between knowledge (when defined in terms of these concepts/ideas), and ethics, morality, and religion. The road from Descartes, Kant, and Spinosa to Wittgenstein and Chomsky is unhindered by the uses of these items of knowledge. They can be used to strengthen one's deep religious beliefs or to reinforce a political or social agenda. The phenomenon is the same for the religiosity of Descartes and Kant, and the political activism of Noam Chomsky.

The Search for Nuggets

When information theory and linguistics in knowledge began to converge, we found ourselves with a unit of knowledge that consists of concepts and propositions. One of the units of knowledge that emerged was the "nugget." In Rubenstein and Geisler (2003) we defined a nugget as the "units that carry the knowledge" (p. 2). We also distinguished between "intellectual nuggets," "supernuggets," and "nugget events."

Intellectual nuggets are defined as the basic unit of knowledge. These are compound statements, some in the form of *causal statements,* others suggesting *correlations.* Supernuggets are a construct related to an organization's attempt to build a knowledge system. They are a set of nuggets that are "tied specifically to expressed or implied needs of the organization" (p. 2).

Thirdly, a "nugget event" is defined as "The identification of one or more nuggets that might go into the nugget inventory or pool for current or future use by the originator/identifier or others in the organization" (p. 3).

Nuggets are simply statements or propositions that contain some form of knowledge. As shown in Table 1, some may be true, reflect potential action, ethical or not, or reflecting some human or organizational competence. But, regardless of their attributes, nuggets are statements or propositions describing constructs, concepts, and complex notions through the use of languages. They are not—nor were they designed to be—the basic units of knowledge. Nuggets are established to capture concepts and ideas by using linguistic tools, but they are not geared to measure or to describe the basic elements that form the structure of knowledge.[2]

However, in Rubenstein and Geisler (2003) we did show that "nuggets" are a viable form of knowledge definition that allow people in organizations to create, store, transfer, and otherwise transact in these units of knowledge. Granted that nuggets represent complex concepts or notions, they do however permit a good pattern of communication among people. Nuggets are mechanisms that condense thoughts and ideas into a format that is understandable by, and exchangeable among, humans. Nuggets serve as receptacles for experience, wisdom, and other manifestations of knowledge that people desire, and are able to create and to communicate.

Forms of Knowledge

Before there were "memes," there were ideas, symbols, and myths. These are concepts or constructs that have been defined and described as modes of knowledge. Mircea Eliade (1991) defined myths as narratives of a sacred history: "It relates an event that took place in primordial time." In Eliade's definition, myths tell us how "reality came into existence." Myths are a form of knowledge embedded in the culture of people where realities are described in narratives and transferred across generations.[3]

Paul Feyerabend, in his book *Farewell to Reason* (1988), described several structural arrangements that he called *forms of knowledge*. Among them are *lists* of words that depicted in antiquity the classification of people and economic assets and activities, thus facilitating communication and the written power of language as a sound, political, and economic tool. Other forms of knowledge included *stories* and *accounts,* as well as *histories* of complex events.

A well-publicized attempt to offer some unity in what constitutes "knowledge" was Edward Wilson's concept of *consilience.* In his book by this title (1998), Wilson defines consilience as the interlocking of causal explanation across disciplines so that "an induction obtained from one class of facts coincides with an induction obtained from another different class" (pp. 8-9). Wilson argued for the intellectual intersection among such diverse professional knowledge pursuits as biology, social science, environmental policy, and ethics. He believed that the "great branches of learning" can be unified, so that the human adventure or quest would lead from the "genes to culture": from biology and the physical sciences to the social sciences and their policy and political implications.

The Search for the Unity of Knowledge

With the exploration of the nature of science and scientific inquiry, we also inherited the long-standing effort to search for the unity of knowledge (Bunge, 2003). In addition to Wilson (1998) and Damasio et al. (2001), social scientists and economists have engaged in the pursuit of a framework that would unify their disciplinary and other knowledge. Brinkman and Brinkman (2006) examined the "dichotomy of useful knowledge" in which the physical or "hard" sciences are joined together with the "soft" sciences of the social disciplines and the humanities. They argued that such synthesis of knowledge manifests itself in human *culture*, "…in serving as a storage or reservoir of society's accumulated knowledge, provides a blueprint or code for humankind's social life. In this regard culture serves as humankind's social DNA" (Brinkman, 1981, p. 107).

The notion of culture as a repository of human knowledge is postulated upon at least two assumptions. The first is the ability of humankind's culture to evolve and to continually accept and incorporate new knowledge. Brinkman and Brinkman (2006) refer to this: "Culture evolution enables humankind to take bigger and bigger bites of the infinity of knowledge stored in the core of culture" (p. 448).

The second assumption is the existence of adequate mechanisms, benchmarks, and criteria for the absorption of new knowledge into the "core" of culture. This assumption is conditioned upon the ability of the scientific disciplines (both "hard" and "soft") to come together in a synthesis that overcomes the disciplinary divides.

This last condition is obviously easy to state but very difficult to satisfy. As Thomas Kuhn (1966), Imre Lakatos (1999), and others have suggested, scientific inquiry is not a purely objective endeavor, but a social activity, conducted by people who are influenced by human frailties and who therefore also have subjective and irrational motivation in their drive for scientific exploration.

The entrenchment and isolation of scientific discipline has long been a well-studied phenomenon. Interdisciplinary inquiries are few and difficult to accomplish. Even if such "getting together" does occur, there are few, if any, specific rules, benchmarks, and criteria for such synthesis—precisely the type of instruments needed to absorb diverse modes of knowledge within the human culture (Leary, 1955; Tynjala, 1999).

In recent years some economists have argued that knowledge is the new asset of wealth creation. Such conceptualization of knowledge would require some unity in the types and modalities of knowledge, perhaps "useful knowledge," from many different societies. But, how do we put together such different outputs of scientific inquiry, and what methods and criteria do we then employ to create a workable synthesis? How do we assimilate and synthesize findings from an experiment in physical chemistry with findings from a geological survey, a study of human preferences for modes of transportation, and perhaps also combined with results from a study of genetic abnormalities in the formation of the retina in mice?

Even though we may embrace a pecuniary approach to the value of knowledge as it is synthesized from various disciplines, the lack of tools for such a combination to successfully happen and the inherent barriers in the cross-disciplinary fusion are insurmountable factors that condemn such an effort to failure. In the search for unity of knowledge, there has also been an underlying assumption that the similarities between the disciplines of scientific inquiry far outweigh the differences, so that necessary principles of cohesion and tools and mechanisms for synthesis are bound to emerge. Those who joined the search for unity of knowledge tended to favor the complexity of culture and the evolutionary model expressed in genetics, for the environments in which the binding of different types and sources of knowledge can take place. Yet, in all of this exploratory effort, there was little, if any, attention given to the basic unit of knowledge, what it is, and how it can be measured.

Meme: The Hype and the Continuing Search

Wilson also argued for the process or phenomenon of co-evolution of genetics and culture, transporting principles of genetics to the cultural concepts and the social sciences upon which they are based. He advanced the unit of culture as the "node of semantic memory and its correlates in brain activity. The level of the node whether concept (the simplest recognizable unit), proposition, or schema, determines the complexity of the idea, behavior, or artifact that it helps to sustain in the culture at large" (p. 148). As Wilson concedes, this is another definition for the *meme,* the term coined by Richard Dawkins in 1976.

Dawkins, a professor at Oxford University, coined the term *meme,* proposing that this is the unit of information or knowledge that acts in the social or cultural environment in a way similar to that of biological genes. Memes are the basic elements of culture. They replicate themselves and move from brain to brain as humans transact in information and knowledge. In addition to such processes of replications, memes evolve (in a way similar to that of biological evolution) by not only serving as carriers of knowledge across people, but also by modifying this knowledge in an evolutionary mode by learning and adapting—as do biological entities.

Memes are therefore ideas, concepts, and the like to which Dawkins and his followers attribute characteristics found in genes, even in biological organisms, as are viruses, which infect other organisms, replicate, and adapt to changes in their environment.[4] These "memetists" ascribe such characteristics to the processes by which ideas proliferate in society. For example, the proliferation, transfer, and absorption of complex concepts such as "private enterprise," "freedom," "democracy," "the rule of law," and even "civilization" and "human rights" are viewed as social equivalents to viral infestations.

This "Oxford School" of Memetics (following in the footsteps of Dawkins, and having the Oxford University Press publish many of the books on this topic) also has adherents in the United States. In 1997 they founded a dedicated electronic journal: *The Journal of Memetics—Evolutionary Models of Information Transmission.*[5] Moreover, scholars in this area have recently argued that memetics is a new and formal science. Bruce Edmonds commented that: "Memetics is, at the moment, a gloriously diverse field—it ranges all the way from a narrative framework in fields such as history and anthropology to formal predictive theories in biology and computer science...the application of models with an evolutionary or genetic *structure* to the *domain* of (cultural) information transmission" (Edmonds, 1998). Edmonds (2002) also advanced three challenges to memetics: a conclusive case study, a theory of memetic modeling, and the emergence of a memetic process.

Lissack (2004) proposed a redefinition of the meme. He argued that the field of memetics has overly focused on the nature or structure (ontology) of the memes-as-beings, in the tradition that they are equal to genes in the mode of their propagation in the cultural environment. Instead, Lissack suggested that memes should be:

"...a label for successful boundary object indexicals and lose their privileged status as replicators. Instead, the replicator status is ascribed to the environmental niches and the memes are their representatives, symbols, or semantic indexicals. With this definition, memes are repackaged as symbols

and their impact or management is not that of a viral contagion but rather as an indicator of success and change in environmental niches."

I have devoted considerable space to the search for memes, not only to describe the intellectual gymnastics in the creation of this concept—but also to build the necessary foundation before the inevitable critique of this effort and the conclusion, in the following pages, that the search for structure does indeed continue unhindered. More on memetics will be discussed in the second part of this book, where I include memetics in the criticism directed at evolutionary theories of the progress of knowledge.

Here, however, suffice it to say that "memes"—however redefined and packaged—are a poor descriptor of the basic elements in the structure of knowledge. However memeticists manipulate this concept, memes are still linguistic representations of ideas or concepts that have been reified and endowed with ontological prowess of independent action.

A similar description of the elemental particle of knowledge is the "granule." This construct is defined as a non-empty set of objects that contain information and has a non-empty set of attributes. For example, Wang and Wu (2003) suggested that "the information system (IS) may be divided or covered by the set consisting of these granules, which gives an approximation to the IS and can be named a granule view of it." They also added that "the granule view of an information system depends on the similarity used to form a granule." These similarities they describe may be equivalence, tolerance, or reflective-binary relations.

"Granulation" of the element of knowledge offers some interesting avenues in the definition of particulates of knowledge. They are, however, only a first approximation to the unit of knowledge as I describe it in this book. Without a more stringent and detailed description of what they are and their constituent elements, granules add little beyond a fractal fragmentation of the structure of knowledge.

The attempts to define the elements of knowledge as granules differ from the effort to describe them as memes. The former is an exercise in slicing the concept of knowledge into fractions. These are essentially identical to the larger concept, but are structurally smaller and simpler. Memes, on the other hand, are the result of a more sophisticated approach, linking evolutionary theory and social uses of it (Kitcher, 2004). I will not engage in a debate on the validity of applying evolutionary theory to social uses or the fallacies of sociobiology. As Kitcher (2004) has shown, Gould (2002) and Gould and Lewontin (1993) have offered strong critique, which I gladly support.[6]

Supporters of memetics tend to argue that the cultural replicator is not the "idea" but the mind, or the view of the world within the mind. Gabora (2004) argued: "An idea participates in the evolution of culture by revealing certain aspects of the worldview that generates it, thereby affecting the worldviews of those exposed to it. If an idea influences a seemingly unrelated field this does not mean that separate cultural lineages are contaminating one another, because it is worldviews, not ideas, that are the basic unit of cultural evolution" (p. 127). Clearly, the road from basic unit of culture to basic unit of knowledge is short and almost inevitable. However knowledge may be diffused, in networks of individual brains or via networks of machines and human interacting, the unit of knowledge must be constructed in a form that complies with structural exigencies of what is the elemental unit. We must distinguish between the structure of knowledge and its growth, diffusion, and progress. Memes are not building blocks of knowledge. Whether ideas, models, or worldviews as reflections of reality, they are ontological constructs that may contain knowledge but do not represent, in and by themselves, the basic unit of what constitutes knowledge. To use a cliché, they are the context for knowledge, not the content of it.

Categories, Attributes, and the Search for the Basic Unit

One of the early scholars whose model of knowledge approximates my model of the unit of knowledge was Werkmeister (1948). In a chapter on *The World About Us,* he discussed human experience and the human analysis of the external world by distinguishing among categories of "otherness." He argued: "It is evident, therefore, that the well-differentiated manifoldness of 'otherness', positively interpreted, leads to specific sequences or *dimensions* and, therefore, to specific elements of order in 'my' first-person experience" (pp. 86-87).

Werkmeister also suggested that experience is based on the realization that *sensory qualities* (as he called them), such as colors and sounds, are different from *other* mental images or hallucinations. He proposed that "the 'dimensions of otherness' previously referred to are now found to be interrelated in such a way that they constitute large 'realms of otherness.' The whole of experience is thus permeated with these elements of order" (p. 87).

Since the framework proposed by Kant, this is the first reference to such categories and their "inter-relationships" to form larger concepts that I found in the literature and that systematically refer to the nature and structure of knowledge. Werkmeister proposed seven pairs of such basic categories:

1.   Quality-Quantity
2.   Unity-Manifoldness
3.   Form-Matter
4.   Universal-Particular
5.   Relation-Substratum
6.   Dimension-Opposites
7.   Continuity-Discreteness

These pairs of categories are a mix of distinct opposites in continua of a concept (such as "continuity-discreteness") and some categories that are different aspects or attributes of an experiential event (such as "quality-quantity" and "form-matter").

Following in the tradition of Kant's categories, Werkmeister proposed that the categories of the first-person experience "provide the basic elements of order which make an integration and interpretation of that experience possible." But, he added the elements of *temporality* and *spatiality,* so that the interpretation of the categories of "otherness" can be ordered as to their relation to time and space.

In his discussion of the nature of the world outside the "self," Werkmeister has raised several issues that are not entirely resolved in his book. He talked about "configurational complexes of qualities which move together as a unit," and "configurational patterns" of such qualities. He also mentioned the issue of different dimensions of the "otherness" (in the world external to the self) that sensory qualities seem to represent. However, he stopped short of exploiting these insights and progressing into a cohesive model of the basic unit of knowledge.

In summary, the search continues for the basic unit of knowledge. Several fundamental issues were brought to the fore by writers who had joined the search: How do we "really" know the physical world about us? Does this world "truly" exist as an entity separate from ourselves?" And what causes the 'integration' of the *qualities* we assign to inputs from the external world?

Such philosophical queries may have unwittingly diverted the attention of scholars from the search for the unit of knowledge to concentrate on their pursuit of epistemology as a means to explain the knower's view of the external world. Perhaps this was done because the search for the unit of knowledge is more a methodological adventure, whereas understanding how we view the "world about us" is more conceptual, hence much more attractive to the curious intellect (Kaplan, 2006).

The third question listed above (what causes integration of qualities), discussed in Werkmeister (1948), is a crucial element of the model of the basic unit of knowledge I present in Chapter IV. I return to these various methodological and ontological problems (which emerge when we attempt to "integrate" or cluster qualities of externally derived inputs) upon the discussion of my model.[7]

Levels of Inquiry

The relatively puny searches for the basic unit of knowledge and the much more intensive study of knowledge in epistemic terms are two levels of inquiry aiming at a similar objective: to understand the nature of human knowledge and to be able to measure it. This dual path to the investigation of knowledge is composed of two levels. The first is the level of the basic unit; the second, the overall view of the phenomenon.

In many ways this is similar to the duality in the investigation of matter and physical phenomena. There is an established incompatibility between nature's behavior under the General theory of relativity (which aims to explain the macro world of matter) and quantum mechanics, which deals with the behavior of sub-atomic particles (e.g., Sachs, 1988). Distinctions between the two theories or approaches include conceptual, as well as methodological or mathematical differences. In relativity theory, for example, space-time is a continuum, with large distances between aggregates of matter, whereas in quantum mechanics, the very small scale of the elements of matter requires non-linear mathematics, indeterminism, and other modes of investigation which do not apply to the larger world of physics.

In 1935, Erwin Schrödinger (1887-1961), an Austrian physicist, wrote a letter to Albert Einstein, then published a paper which provides an example of the measurement problems in quantum mechanics (Schrödinger, 1935; also see Gribbin, 1984). The example is known as "Schrödinger's cat," and its purpose was to illustrate the differences between relativity and quantum mechanics. Schrödinger suggested:

"...one can even set up quite ridiculous cases. A cat is penned up in a steel chamber, along with the following diabolical device (which must be secured against direct interference by the cat): in a Geiger counter there is a tiny bit of radioactive substance, so small that perhaps *in the course of one hour* one *of the atoms decays, but also with equal probability, perhaps none; if it happens, the counter tube discharges and through a relay releases a hammer which shatters a small flask of hydrocyanic acid. If one has left this entire system to itself for an hour, one would say that the cat still lives if meanwhile no atom has decayed. The first atomic decay would have poisoned it. The psi function for the entire system would express this by having in it the living and the dead cat mixed or smeared out in equal parts."*

In his example, Schrödinger pointed to the case where events of a very small phenomenon with several possible outcomes would become a clear, deterministic description of a macro-reality. He wrote: "It is typical of these cases that an indeterminacy originally restricted to the atomic domain becomes transformed into macroscopic indeterminacy, which can then be resolved by direct observation. This prevents us from so naively accepting as valid a blurred model for representing reality."[8]

Searching for Unity and the Theory of Everything

The conflict between the "macro" and the "micro" perspectives of the physical world has ignited a search for a theory whose purpose would be to unify the two perspectives. Such a theory would also be a "theory of everything" that exists: in the cosmos and inside the atom.[9]

This search has led to the development of the "String" or "Superstring" Theory. In a singular effort which combined the basics of music and problems of physics, this theory was conceived. It proposes a basic structure of matter, at this point beyond the ability of human experience and measurement tools (e.g., Davies & Brown, 2006; Greene, 2000). Matter, hence also the universe, is made of infinitesimally small "strings"—which are vibrating loops of energy—in a manner similar to that of strings in musical instruments. Such vibrations of strings and the combination of vibrating strings form different "realities" or dimensions, which are beyond human grasp, due to the limitations of our senses and our instruments.

The challenges of the string theory, particularly in the mathematical horizons it opens for exploration, have attracted many scholars to this exciting new field of inquiry, as well as many detractors (see Greene, 2000; Ward, 2002; Wheatly, 2001). There seems to be an immensely powerful drive and an aphrodisiac attraction in the will to arrive at a theory that can explain everything: from the strong forces such as gravity to the weak forces such as those holding together the sub-atomic particles.

The British astrophysicist Stephen Hawking has also aptly articulated this drive in his effort to explain cosmological anomalies such as "black holes." Hawking has incessantly engaged in what he believes has been the uninterrupted search by a host of scholars who preceded him to uncover a theory that would "unify" or explain all the various forces operating in the universe, *at all levels.* Obviously, such a research effort leads to the search for the *very* basic or elemental unit of whatever we are researching or trying to explain (Hawking, 2006).[10]

In the case of the cosmos and the nature of matter, it would be going below the level of sub-atomic particles. Ever since the Greek philosopher Democritus (circa 460 BCE-370 BCE) coined the term "atom" to describe the indivisible element of all matter, there has been a quest to discover the elemental components of even this "indivisible" unit.

This "perfect drive" has also been the motivation behind this book and this author's quest for the elemental unit of knowledge. The differences between the problems of knowledge itself[11] and those of the macro-approach to knowledge (such as managerial and organizational knowledge systems)[12] call for a framework that would unify these diverse areas of investigation. Hence there is also here the need to explore the *very* basic unit of knowledge.

References

Aunger, R. (2002). *The electric meme: A new theory of how we think.* New York: The Free Press.

Bergson, H. (1998). *Creative evolution.* Mineola, NY: Dover.

Brinkman, R. (1981). *Cultural economics.* Portland, OR: The Hopi Press.

Brinkman, R., & Brinckman, J. (2006). Toward a grand union: The banyan tree of knowledge. *Journal of Economic Issues, 40*(2), 439-448.

Bunge, M. (2003). *Emergence and convergence: Qualitative novelty and the unity of knowledge.* Toronto: The University of Toronto Press.

Cadamer, H. (1977). *Philosophical hermeneutics.* Berkeley: University of California Press.

Damasio, A. et al. (Ed.). (2001). *Unity of knowledge: The convergence of natural and human science.* New York: New York Academy of Sciences.

Davies, P., & Brown, J. (Eds.). (2006). *Superstrings: A theory of everything.* New York: Cambridge University Press.

Dewey, J. (1977). *How we think.* Mineola, NY: Dover.

Durham, W. (1991). *Coevolution: Genes, culture and human diversity.* Stanford, CA: Stanford University Press.

Edmonds, B. (2002). Three challenges to the survival of memetics. *Journal of Memetics, 6*(1).

Edmonds, B. (1998). On modeling in memetics. *Journal of Memetics, 2.*

Eliade, M. (1991). *The myth of the eternal return.* Princeton, NJ: Princeton University Press.

Ferguson, K. (1992). *Stephen Hawking: A quest for the theory of everything* (reprint ed.). New York: Bantam Books.

Feyerabend, P. (1988). *Farewell to reason.* New York: Verso Books.

Gabora, L. (2004). Ideas are not replicators but minds are. *Biology and Philosophy, 19*(2), 127-143.

Gould, S.J. (2002). *The structure of evolutionary theory.* Cambridge, MA: Belknap Press.

Gould, S.J., & Lewontin, R. (1993). The spandrels of San Marco and the Panglossian paradigm: A critique of the adaptationist program. In E. Sober (Ed.), *Conceptual problems in evolutionary theory.* Cambridge, MA: MIT Press.

Greene, B. (2000). *The elegant universe: Superstrings, hidden dimensions, and the quest for the ultimate theory.* New York: Vintage.

Gribbin, J. (1984). *In search of Schroedinger's cat: The startling world of quantum physics explained.* London: Wildwood House.

Gribbin, J. (2000). *The search for superstring, symmetry, and the theory of everything.* Boston: Back Bay Books.

Halpern, P. (2004). *The great beyond: Higher dimensions, parallel universes and the extraordinary search for a theory of everything.* New York: John Wiley & Sons.

Harman, G. (2002). Reflections on knowledge and its limits. *The Philosophical Review, 111*(3), 417-428.

Hawking, S. (2006). *The theory of everything: The origin and fate of the universe.* San Francisco: Phoenix Books.

Kaplan, B. (2006). Deploying a knowledge convergence framework. *Knowledge Management Review, 9*(3), 18-21.

Kitcher, P. (2004). Evolutionary theory and the social uses of biology. *Biology and Philosophy, 19*(1), 1-15.

Kuhn, T. (1966). *The structure of scientific revolutions.* Chicago: University of Chicago Press.

Lakatos, I. (1999). *Proofs and refutations: The logic of mathematical discovery.* New York: Cambridge University Press.

Leary, L. (1955). *The unity of knowledge.* New York: Doubleday.

Leydesdorff, L. (2001). *A sociological theory of communication: The self organization of the knowledge-based society.* Boca Raton, FL: Universal.

Lissack, M. (2004). The redefinition of memes: Ascribing meaning to an empty cliché. *Journal of Memetics, 8*(1).

Rubenstein, A.H., & Geisler, E. (2003). *Installing and managing workable knowledge management systems.* Westport, CT: Praeger.

Sachs, M. (1988). *Einstein versus Bohr: The continuing controversies in physics.* Pern, IL: Open Court.

Schroedinger, E. (1983). Die Gegenwartige situation in der quantenmechank. In A. Wheeler & W. Zurek (Eds.), *Quantum theory and measurement* (pp. 807-812, 823-823, 844-849). Princeton, NJ: Princeton University Press.

Steel, D. (2004). Social mechanisms and causal inference. *Philosophy of the Social Sciences, 34*(1), 55-78.

Thow-Yick, L. (1998). General information theory: Some macroscopic dynamics of the human thinking systems. *Information Processing & Management, 34*(2/3), 275-290.

Tynjala, P. (1999). Towards expert knowledge? A comparison between a constructivist and a traditional learning environment in the university. *International Journal of Educational Research, 31*(2), 357-442.

Wang, L.-H., & Wu, G. (2003). Attribute reduction and information granularity. *Journal of Systemics, Cybernetics and Informatics, 1*(1), 36-42.

Ward, M. (2002). *Beyond chaos: The underlying theory behind life: The universe and everything.* New York: Thomas Dunne Books.

Werkmeister, W. (1948). *The basis and structure of knowledge.* New York: Harper & Bros.

Wheatly, J. (2001). *The nature of consciousness: The structure of reality: Theory of everything: Scientific verification and proof of logic God is.* Asland, OH: Research Scientific Press.

Williamson, T. (1999). On the structure of higher-order vagueness. *Mind, 108*(429), 127-143.

Wilson, E. (1998). *Consilience: The unity of knowledge.* New York: A. Knopf.

Wolin, S. (2004). *Politics and vision: Continuity and innovation in western political thought.* Princeton, NJ: Princeton University Press.

Endnotes

[1]   See: Hirschberg, S. (2002). *Past to present ideas that changed our world.* Upper Saddle River, NJ: Prentice Hall; Berlin, I., & Hardy, H. (Eds.). (2001). *Against the current: Essays in the history of ideas.* Princeton, NJ: Princeton University Press; Fernandez-Armesto, F. (2003). *Ideas*

that changed the world. New York: DK Publishing; Yergin, D., & Stanislaw, J. (2002). *The commanding heights.* New York: The Free Press. In this book, the authors describe the role that free trade and economic opportunity—as captivating ideas—changed the way most world economies are managed today and how these ideas contributed to the fall of communism and the Soviet Union. Also see Wolin (2004) for a description of political ideas. Another example is Mandelbaum, M. (2002). *The ideas that conquered the world.* New York: Public Affairs Publishers. Mandelbaum discusses peace, democracy, and the triumph of free market ideology as the pillars of the twenty-first century.

2   Nuggets are very useful in studying knowledge systems in organizations. Rubenstein and Geisler (2003) have shown the role that intellectual nuggets play in the design and implementation of organizational knowledge management systems (KMSs).

3   Myths can also be regarded as *clusters* of inputs, so that they are structures similar to that of any form of knowledge I describe in Chapter IV.

4   See, for example, Aunger (2002) and his discussion of the Electric Meme that supports the notion of replication of cultural ideas. Also see: Brodie, R. (1996). *Virus of the mind: The new science of the meme.* Seattle: Integral Press; Dawkins (1976); Blackmore, S. (1999). *The meme machine.* Oxford: Oxford University Press; Aunger, R. (Ed.). (2000). *Darwinizing culture: The status of memetics as a science.* Oxford: Oxford University Press; Jefferys, M. (2000). The meme metaphor. *Perspective in Biology and Medicine, 43*(2), 227-242.

5   See *http://jom-emit.cfpm.org*

6   Kitcher (2004) both praised Gould's contributions to evolutionary theory early in Gould's career, and criticized his later essays in which Gould was unhappy with the extension of evolution theory to sociobiology. Kitcher writes: "Darwin gave us a metaphor, the image of natural selection. Now the breeder, interested in a particular property of the flower or the pigeon, does select for a particular trait. Nature doesn't." This statement encapsulates the essence of the critique of transferring biological evolutionary ideas to the social milieu where nature acts at a randomized will, social events are much more manipulable: human rather than natural. This is even more paramount in the conception of knowledge.

[7]   It is difficult to isolate the search for the basic unit of knowledge from the general problems of epistemology. In doing so, I will probably be ardently criticized by philosophers of almost every specialty. Arguments will probably be advanced in which unresolved problems in epistemology will be an excellent surrogate for direct critique of the model itself. See such possible arguments in Steel (2004), Durham (1991), Aunger (2001), and Thow-Yick (1998). Nonetheless, I am convinced that the *separation* of the search for the basic unit from the general issues of epistemology and ontology is not only possible and feasible, but also necessary. Such a search brings a fresh outlook to this ancient quest, and its outcome may even serve as a guide for the direction to be adopted by those who seek larger philosophical pursuits.

[8]   This is also akin to the "Sorites Paradox," which I discuss later in this book. In the effort to link the elemental unit of knowledge to the formation of "macro" constructs of knowledge such as concepts (through the action of cumulation), we encounter a similar problem. There are several possible states, in the form of "superposition," and we are confronted with the "Sorites" problem: when will the cumulation of "individualized" or microscopic elements of knowledge combine to such a degree that they form a macroscopic description of a construct of knowledge.

[9]   See, for example, the popular accounts of this search in Gribbin (2000) and Halpern (2004).

[10]   Also see Kitty Ferguson's account of Hawking's search in Ferguson (1992).

[11]   For example, the distinction between tacit and explicit knowledge, and the conjuring of knowledge via the senses. There is also an *intermediate* set of issues regarding communication of knowledge.

[12]   For example, issues of manipulation, access, acquisition, and interpretation of knowledge in such systems (Leydesdorff, 2001).

Chapter IV

Crucible of Synthesis:
The Model of Knowledge, from Sensorial Signals to Architectures and Concepts

Why is the search for the basic unit of knowledge possible and necessary, even when *separated* from larger philosophical and scientific questions? There are three main reasons. First, one such problem deals with the *process* by which knowledge is gained about the external world and rational beings interact with their external environment. A second problem examines the existence or *ontology* of this external world: is it real or an artifact of the mind? Thirdly, another area of inquiry concerns the degree to which we can trust our senses. Are the inputs we receive from the external world "true" representations of the external reality?

In any of these instances, the basic unit of knowledge is a component of the mechanism of search and inquiry. Whatever the unit we determine to be the basic element of knowledge, its being in itself would not impact the theoretical foundations of the inquiry into larger more complex problems, to the extent that they would be refuted.[1]

A model of knowledge based on the elemental unit of knowledge would be later linked to the issues involved with the macro approach to knowledge. Figure 1 shows the three main components of the search for the nature and progress of knowledge.

The conceptual as well as temporal distance between the micro approach (the basic unit) and applications in the macro world of knowledge systems are mediated by the use of modes and mechanisms of linkage or exchange. There are tools of communication, semantics, language, and semiotics. Human interaction and human society and its survival depend on the effective utilization of these mechanisms for the exchange of knowledge—from its basic unit to the complexity of systems.

As in physics, perhaps the rules or principles present within the *micro* perspective may be different from those that apply to the *macro* perspective. Polanyi (1966) and scholars who followed tend to differentiate between the two perspectives primarily in terms of "tacit" and "explicit" knowledge. But, the differences are more salient and complex than simply these two categories. There are profound distinctions in the nature of knowledge, so that its transfer from the individual to higher-order systems becomes a conceptual and structural endeavor of low probability and extreme difficulty, hence the crucial role that the mediating modes of communication play in this phenomenon.

In order to understand these critical differences, it becomes necessary to explore the nature of knowledge and its elemental building blocks. This is similar to nuclear physics to delving into the secrets of sub-atomic particles and their constituent elements.

Basic Elements of Knowledge

The model proposed in these pages is based on two complementary approaches. The first introduces the question: What constitutes the fundamental components of what we know? The second deals with the process and method by which such fundamental components may form a more complex or an initial construct of knowledge. This process or method is not yet an exploration of the *progress* of knowledge. Rather, it focuses on the composition of the structure of knowledge: from the most fundamental unit we can envision, and perhaps measure, to the more complex component of the structure of knowledge.

In this sense, the model I introduce here is similar to the model of the atom and its sub-atomic particles. The structure of knowledge I propose will have at least two levels of aggregation of its most fundamental components which we can conceptualize, as sub-atomic units form the atom, and atoms form molecules.

At the outset it should be established that the fundamental building blocks of knowledge are inherent in the process by which intelligent beings acquire and process knowledge about the world in which they exist. At this juncture there is not a conceptual distinction between data, information, knowledge, and wisdom. Such a hierarchy is an artificial superstructure imposed on the elemental particles of that which tells us about the external world and to which I refer here in generic terms as *knowledge*.

In the case of humans, the fundamental building blocks of knowledge they absorb from their surroundings are *signals* perceptible to human senses. The mental processes by which such sensorial inputs are digested and manipulated have long been discussed in philosophical inquiries as the working of

Figure 1. The main components of a model of knowledge

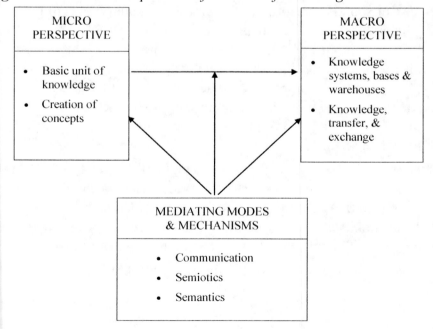

reason, or the human ability to cognitively work with notions, concepts, and ideas. These mental processes will be further examined in Chapter VI and in Part II of this book upon the description of the model by which knowledge progresses. At the very basic process of human aggregation (or clustering) of knowledge, the taxonomical effort of the mind is the initial mode by which sensorial inputs are classified. But, in the beginning of it all, there are signals emanating from the world that we inhabit, and these signals are captured by our senses to ultimately form a mode of understanding of such world—a mode we call *knowledge*.

Signals and Their Attributes

The model of the structure of knowledge I present in this book suggests that the initial exposure of human cognition to knowledge is by the absorption of signals from the environment. This is done by five senses: *hearing, sight, touch, smell,* and *taste.* The senses collect and absorb the following types of signals: *sound, light, temperature, odor,* and *flavor.*

The five senses that humans possess allow them to process signals from their surroundings. These signals are captured and translated into sensations. The translation or processing of raw signals into sensations such as light, flavor, or sound is performed at the most elementary level of processing, very likely by chemical exchanges between the senses and the relevant part of the human brain where those signals are processed.[2] At this level we should not assume the existence of pre-established schemes of classification of the signals received and processed. The reason is that there has not been an operation of clustering or classification into distinct categories. Hence, each of the senses performs its own processing of its specific signal: vision, touch, smell, and so forth.

To create meaning or assign a superimposed rationale on the signal received, it is necessary to have attributes that will serve as criteria for classification and clustering. But, to have this pool of signals and their attributes, we must start with the pool itself—the signals as they are received and *acknowledged* by the senses—*before* they are classified and interpreted.[3]

These signals have one or more of the following attributes:

- Form/shape (size, hardness, texture)
- Speed (movement)
- Space (location/position)
- Time (mobility)
- Distance (from a given point in space)
- Direction (of movement)
- Intensity (of signal)
- Measurability
- Clarity of signal
- Life or continuance (how long the signal lasts)
- Connectivity or relation to other signals

The attributes of these signals are the characteristics of the signals (or inputs from the environment outside the receiver of these signals). The following questions arise: How are these attributes recognized by the human mind (the receiver of signals)? Are these attributes inherent in the signals, or the product of the mind's cognitive process? And, are these attributes *categories* by which the mind classifies these characteristics of the signals from the external world?[4]

The attributes do reside in our mind and serve as criteria by which occurs the first-step manipulation of the inputs from our senses. They are not qualities of the "real" things in the external world being surveyed by the senses. They are the artifacts or instruments by which the mind introduces some measure of *order* into the flow of signals it receives. These are not yet the conceptual categories Kant had proposed.[5] They are nonetheless one stage prior to the rational construction of meaning that would fit the Kantian category.

Consider, for example, the vision of a geometric shape such as a triangle. This geometric shape also emits an auditory sound and is cold to the touch. The three inputs of sensorial signals are transmitted via the senses to the mind (or to a central location or processing unit). They are *descriptors* of the geometric shape encountered in the environment. In their appearance, these inputs are sensorial contingencies that convey to the mind distinct descriptions of a *unique* occurrence. Whether it is an object or another element of the external world remains to be seen.

Physicalism and Qualia

The question I posed earlier: "How are these attributes recognized by the human mind?" has been the basis for a long-standing contention among philosophers of the mind, philosophers of science, behavioral scientists, and many other scientists interested in this topic (Harnad, 2000; Humphrey, 2000; Rose, 1999).

Two major schools have emerged, holding opposing perspectives on what might constitute a plausible answer to this basic question of the nature of knowledge. One such school has been termed: "Physicalism" (Quine, 1977). Its main argument in this reference is that everything is physical—that is, everything in nature is physical or material. Although this perspective was meant to describe nature and the universe, it has also been applied to the nature of knowledge and the processes of the mind (Savire & Kandel, 2000). Physicalism then advanced the argument that mental events or phenomena can be reduced to physical properties and processes. For example, the recognition of attributes of sensorial inputs would then be reduced to physical processes, such as chemical reactions, or even sub-cellular processes. Proponents of the physicalism perspective have advanced several arguments for it, and have provided different modes of physicalism such as "minimal physicalism," which attempts to arrive at the core ideas of this perspective.

Such arguments in favor of physicalism have not been very convincing. This perspective also led to the emergence of arguments against it. There are at least two key categories of counter-arguments or perspectives that defy physicalism as it applies to mental processes and knowledge creation in the mind: intentionality and qualia. Intentionality refers to the argument that thinking or knowledge is *about* something, so that the mind thinks intentionally of an object or a notion. This argument is fuzzy and elevates the discussion to levels of conceptual constructs, without convincingly refuting the physicalist perspective (Averill, 2005; Dennett, 1991).

The perspective that advocates the existence of *qualia* has been much more prevalent in the literature opposing physicalism. What, then, are qualia? They are the intrinsic properties or attributes of experiences of senses, such as colors, pleasure, or pain. Another definition is the "qualitative content of mental states" (Dennett, 1991). Qualia are defined as a manifestation of human consciousness, which is either different from the physical process that generated the event or item of experience, or a higher level of abstraction of such a physical occurrence. The preferred example by proponents of the existence of qualia is the experience of color, particularly the color "red."

Beyond the physical process by which signals of light are processed by the human eye to elicit the color red, there is also the experience of "redness," or seeing red, which is experienced by the person seeing red, and is a subjective feeling or experience—separated from the physical process (Quine, 1977). In effect, this is the ultimate tacit knowledge. Dennett (1988) had elucidated this point by suggesting that qualia have four properties that make it utterly impossible to share: (1) ineffable—cannot be communicated, (2) intrinsic, (3) private, and (4) apprehensible in consciousness.

A famous argument in favor of qualia is the "Knowledge Argument" (Jackson, 1982). It contains the description of a hypothetical example in which a person, Mary, is confined from birth to a room that has only the colors *black* and *white*. Mary knows all the *physical* facts about the color experiences of other people. When Mary leaves the room for the first time since birth, she sees the color *red* for the first time in her life. Thus, Mary gains knowledge about the color red that she did not have while in the room—she experiences the color red—hence qualia exist independently from the physical properties of seeing a color (Nida-Rumelin, 1996).

The perspective that qualia exist has been criticized on the basis of cognitive processes and fallacies in the experiment. Churchland (1997), for example, argued that cognitive processes of vision are learned from birth and cannot be learned upon leaving the room. Dennett (1991) argued that one cannot tell whether, upon Mary's leaving the room, there has been a change in qualia.

This substantial literature of arguments and counter-arguments still failed to offer a convincing explanation to how sensorial inputs or experiences form higher-order constructs of knowledge. What if indeed mental processes can be reduced, to an extent, to physical events? As we are continually learning, brain waves and certain areas of the brain seem to "light up" or show activity in connection with certain senses, sensations, even feelings.[6] Human senses are biological functions, hence would be related to and influenced by biochemical processes in the brain. Yet, such a reductionist view still fails to explain how knowledge constructs are formed.

Similarly, what if there is a mechanism, existing independently of the physical structure of the mind, what function is to be the bridge between sensorial inputs and higher-order constructs—all the way to the notion of consciousness? Although there may be some rational arguments to the contrary, is it not possible that these two phenomena may exist simultaneously (Gertler, 1999)?

In the two cases of physicalism and qualia, there is a lack of both theoretical/conceptual foundation for their acceptance and the methodology and instruments for adequate measurement. For example, in the proposed experiment of Mary and her seeing the color red for the first time, there is the possibility that Mary was exposed to more than just the color red. She was also exposed, from birth, to other senses (such as taste and touch—otherwise how would she survive?). Furthermore, there is reductionism not only in the physical perspective of senses and knowledge, but also in the subjective perspective of qualia.

Simply put, experiencing "redness" is not only an experience in color, because color always appears as an attribute of a shape or a form. At least, color is not seen independently of some form. Here one can initiate the reductionist action, whereby shapes (such as a triangle) are reduced to position in space, and to the experiences of distance and size. Thus, isolating an experience such as the color "red" from its other components in human cognition would require reductionist action, which would be limited and not complete in providing the other elements of the total experience of seeing color.

Both proponents and critics of the "Mary" experiment consider the knowledge she has before leaving the room which she may or may not have learned to be a measurable unit—although it is much more complex. What these scholars have identified as *knowledge* is a composite of higher-order constructs, rather than elemental units of analysis. Proponents and detractors alike then proceeded to apply manipulation of logic to what such knowledge would or would not be in this peculiar example.

Scholars have proposed several types of knowledge to explain in their logical manipulation the role that acquisition of knowledge would play in Mary's perception of color. They tended to differentiate between, for example, "knowledge about the physical world," "knowledge as abilities (skills)," and "acquaintance or indexical knowledge (knowing something by becoming acquainted with it, like a city)." In all of these attempts, there is no agreement on the nature of knowledge beyond disagreements on the taxonomical manifestations of what Mary knew and currently knows (Lee, Kageura, & Choi, 2004; Marks, 1978; Pinker, 1994; Small et al., 2006).

These two problems seem to affect the continuing disagreement among scholars on the issue of physicalism and qualia: the definition of what constitutes knowledge in this debate, and the exposure of the mind to the avalanche of sensorial inputs from *multiple* senses simultaneously. This may have clouded the conflicting analyses of the phenomena.

An Alternative Illustration:
The Box of Additive Senses

To illustrate this point, imagine a box containing a robot named "Mario," whose brain has been made to simulate (in all possible details, including the ability to have feelings and human sentiments) the human mind and nervous system. Upon its birth (when a button is pushed and the robot Mario is activated), Mario is exposed to only one of the senses. Subsequently, in random intervals, Mario is exposed to another sense, until he is exposed to all five senses. At each point of cumulative exposure, Mario's knowledge is measured by a series of questions, such as: "What do you see?" or "What do you hear?" and "What is this object?" Simultaneously, in an identical box, Mario's identical twin "Michael" is activated, then exposed to *all* five senses at once. Assuming that we are able to indeed measure the types and levels of knowledge Mario and Michael possess at the conclusion of these experiments, several interesting hypotheses may be tested.

One such hypothesis refers to differences in acquired knowledge to be found in the twin robots. Mario would not be able to "know" a phenomenon with limited exposure to only one sense and its inputs. To experience the color "red" with inputs from vision alone may not be possible without some foundational structures, frameworks, or architectures that are formed by other senses. Hence, one might argue that any knowledge or sensation is based on the cumulative effect of multiple senses. In this case, Michael would have improved facility in creating in his mind knowledge about constructs such as forms and colors, in what may be hypothesized to be a "knowledge explosion," by the cumulative effects of inputs from multiple senses working simultaneously.

The sensorial inputs convey descriptions of this unique occurrence as, for example, form (a triangle), location, speed, and the like. But, for the mind to establish any meaning to the occurrence (such as it being a triangle, of a large size, located nearby, and moving at a given speed), the mind would need to have the pre-established notions of each of the attributes. There would be notions of what is large or small, what is location in space, and what is meant by speed of movement. This would require either a previous experience with such an occurrence (hence the ability to compare between experiences that are temporally apart), or a set of conceptual frames of reference (concepts, categories, or however we call them) that help the mind to classify the sensorial input with a specific meaning of each attribute.[7] This recognition that the senses describe a unique occurrence is not yet knowledge. It is also not a form or mode of clustering. So, what is this stage in the processing of sensorial inputs by the mind?

How and When Cognition Begins

In his insightful research on cognition, Lawrence Barsalou (1999, 2003, 2003a) has genially summarized the contemporary approach to cognition. He described the movement initiated by behavioral and linguistic scientists, in which they distanced themselves from the perceptual view of cognition. Until the twentieth century, cognition (and the structure of knowledge) was considered the result of perception and the creation in the mind of images that represent the external world. In this manner, inputs from our sensors are received, and they form a symbolic representation or a perceptual image of the external reality from which these signals were received. These images (or perceptions of the external world) are stored in memory and serve as frames of reference for any future experiences with the external reality they represent.[8]

More recently, contemporary theories of cognition have rejected, by and large, the notion of perception and symbolic representation of the external world in human memory. Instead, they proposed that cognition works in the following manner. The first step is the formation of perceptual representation from the signals received by the senses. But these perceptions are not then used as images with which to compare and recognize future experiences. Rather, they are "decomposed" into the perceptual components of the external world, then reassembled into higher-order constructs by means of such methods as schemata, semantic, and other logical systems.

In this approach, upon receiving inputs from the senses, the mind does not form a pictorial representation (which mimics the external object as closely as possible), but instead re-enacts reality by reassembling the components of the object that can be derived from the signals. Next, this assembly may be executed with a variety of criteria and systems of inference.[9] Barsalou

(1999) bravely described the problems associated with the contemporary nonperceptual approach. He suggested that a grounding problem affects this approach, so that: "Just as we have no account of how perceptual states become mapped to amodal symbols during transduction, neither do we have an account of how amodal symbols become mapped back to perceptual states and entities in the world" (p. 580). Another problem was identified as the difficulty in having the components of the nonperception system achieve a state of meaning or reason. Barsalou posited that: "Because the processing of amodal symbols is usually assumed to be entirely syntactic (based on form and not meaning), how could such a system have any sense of what its computations are about?" (p. 580).

At this point the reader may believe that I have compromised my promise to minimize the esoteric discussion of these topics. Perhaps this is true. Let me then summarize the topic thus far. Regardless of the differences between the models that suggest perception or conception, both models make what I would call "the contextual leap" from inputs to concepts. This leap calls for going from sensorial inputs received from the environment and from inside the body to fully developed conceptual frameworks, or logical notions. This leap is undertaken over the gap that seems to exist between what we receive as sensorial inputs and how we construe them. There is a need to close this gap.

Some cognitive scientists suggest that there are "zones of convergence" in which sensorial inputs are transformed into images by neural activities. Another aspect of these transformations is the proposed existence of conscious and unconscious modes of processing inputs.[10] Others also argue that conceptualization (with the sensorial inputs) is a process similar to simulation, whereby short-term configuration of inputs is then entered into the long-term memory and becomes useful knowledge.

But, several basic questions remain unanswered: How are simulations formed *for the first time,* with the first imprint? When does *cognition* start? In other words, when do we actually reach the point where sensorial inputs give rise to *knowledge*? Many writers have examined the notions of categories, classes, and representations.[11] There is still a wide gap between such discussions and a useful model that explores the basic unit of knowledge and its formation from elemental sensorial inputs.

In the model I espouse here, the sensorial signals can be analyzed, treated, or classified by the eleven attributes described below. The inputs thus absorbed by the mind are composed of two or more of the *signals* and two or more *attributes* per signal. These are the building blocks of knowledge. The complexity of this initial making of knowledge is the result of the multiplicity of signals and their attributes.

Attributes: Definitions, Types, and Function

Attributes are the criteria by which the mind (or any analytical entity capable of receiving such signals and manipulating them) configures the signals by arranging them in a meaningful format. The attributes-as-criteria can be classified into two major groups: by *context* and by *relations.* The first class is attributes/criteria that describe the characteristics of sensorial signals in relation to the environment whence they had been generated. The classification by relation comprises those attributes that describe the characteristics of these signals as they interact and relate to each other. Thus, the following grouping is possible:

Context Attributes	Relation Attributes
• Speed	• Form
• Space	• Intensity
• Time	• Measureability
• Distance	• Clarity
• Direction	• Life or continuance
	• Connectivity/relation

Attributes-as-criteria can also be classified into three distinct types: (1) concept, (2) function, and (3) tool in the creation of knowledge. As a concept, attributes are categories by which an observer may judge the inputs from the sensors. As a function, attributes are the means toward the assessment of sensorial inputs and the creation of meaning in these inputs. Finally, as a tool in the creation of knowledge, attributes serve as the first step in what I will later describe as clustering of sensorial inputs and the generation of knowledge. Thus, attributes are a tool of the mind or the observable to manipulate sensorial data.

How Images and Experience are Formed

In the formation of images or perception (by manipulation of sensorial data), the roles of experience and memory are crucial. Once sensorial inputs are composed into a meaningful image or concept, such an image (or its equivalent) is deposited in some form of registry or *memory* and an experience of the image or event is imprinted.

There are at least two problems with such a description of how sensorial inputs are transformed and manipulated by a mind or brain capable of manipulating these inputs. First, when was the *first time* such registration of sensorial inputs occurred, and is there a benchmark against which the sensorial input can be compared? Secondly, how does this process of registration and experience formation occur?

In the model of the structure of knowledge I present here, the mind or brain is simply a biochemical plant. It is not a thinking mechanism. Rather, neurons exchange chemicals in a biochemical process in which the *mode* of manipulation of sensorial inputs is the mechanism by which meaning is created. Experience and images (or perceptions) are not imprinted a priori in the mind as categories by which sensorial inputs are assessed and meaning thus attached.[12] The mind is an empty table, on which sensorial inputs are manipulated into a meaningful structure, as if they were the pieces of a puzzle.

So, when is the first time such inputs are manipulated by the mind? Audi (2002), in an analysis of a more complex structure, proposed three modes of memory. The first is a "direct realist" mode, in which the mind remembers an event *as it occurred,* without the mediation of an image it had previously formed of the event. The second is the "representative mode," in which the mind remembers through representation of the event by creating an image or perception from sensorial inputs. Finally, Audi proposes a "phenomenalist view" of memory. This view questions the link between creation of images of an event and the actual act of *remembering.* This third mode leads us into the realm of cognition which is the outcome of the processes described later in this chapter.

For now, suffice it to say that sensorial inputs are collected in the mind and are transformed by applying attributes to these inputs. Some form of an arrangement or architecture is thus formed. This configuration is the first step toward the formation of a structure of knowledge.[13]

The "Big Bang" of Knowledge

If, as I suggest, the mind is an empty table without a priori categories for assessment and manipulation of sensorial inputs, the issues of timing and process duly arise. Timing refers to when the first manipulation appears. Process refers to the *location* of the attributes and *how* the mind applies them without their a priori existence as categories of the mind.

The problem of timing is similar to that of the origin of the universe, long addressed by astronomers. The answer I am proposing is equivalent to the "Big Bang" theory of the origination of the universe as we understand it. Thus, the *"Big Bang" of knowledge* is the first time sensorial inputs are captured by the mind—when the mind is at a stage in which it has the ability to apply or generate attributes that would arrange the inputs in a meaningful format or architecture. This event in the life of the mind may occur at any point in its existence, providing that two pre-conditions are present: (1) ability of the mind to capture and absorb sensorial inputs, and (2) ability of the mind to use attributes to manipulate such inputs—once such attributes are established.

As sensorial inputs first appear and their manipulation occurs, the background is set for the emergence of *experience.* This is the setting of a template in memory, against which other such events may now be compared. It should be noted that this first arrangement of sensorial inputs is unique, in that a specific architecture is formed. It may be the format of a geometrical form (triangle, square, circle) or an object (chair, person, blue sky), or an internal event, such as pain in the digestive system. In this first encounter with the structuring of sensorial inputs, there is no comprehension, nor knowledge, of what these inputs, thus architectured, mean in a larger context of concepts and reason. There is simply a mechanical effort in arranging sensorial in-puts—for the first time—one step above the biochemical exchange among neurons of the brain.

Attributes and Their Characteristics

The attributes by which sensorial inputs are arranged in meaningful architec-tures are not pre-embedded in the brain.[14] They resemble, but are not equal, to the Kantian categories mentioned earlier in this book. Kant's categories are advanced conceptions of the architecture of what constitutes, in Kant's philosophy, the structure of knowledge. The model I describe here considers attributes in a much more fundamental conception of the transformation and manipulation of sensorial inputs.[15]

There are several questions that need to be explored: Where do these attributes reside? How are they recognized by the mind? By which processes are they employed?

The first question is an extension of the assertion that the attributes (or characteristics by which the mind manipulates sensorial inputs) are not embedded in the mind in an a priori manner. The attributes are not a part of the biological or biochemical structure of neurons. They cannot be, because if so, then the question will arise: which came first? If such attributes are embedded in the neurons (as a priori categories of the mind), there would be a logical *exigency* to have neural activity that produced them in the first place.[16]

Nor can we accept arguments that refer to the mind's "natural faculties" (Reid, 1941) or other such conceptual ability, which applies more to the capability of the mind to ascertain truth, sense, and meaning. At this stage I am not concerned with such higher-order constructs, but simply with the manner in which attributes are utilized to arrange and manipulate sensorial inputs.

If the attributes are not embedded in the physical structure of the mind, they are therefore inherent in or native to the environment from which sensorial inputs are received—hence the difference in terms between "attributes" in my model and "categories" in other models. The attributes reside in the environment from which sensorial inputs are generated. The senses are the faculties that allow us to explore the external environment, so that the inputs received are the consequences of such faculties or abilities.

Sensorial inputs, such as vision, hearing, or touch, explore forms or objectives which have certain inherent attributes, such as their context (position in space and time, distance, speed, and direction), and content or relation (such as form). These attributes can be processed in the mind because they (some or all) are the only means by which characterization of the pool of sensorial inputs can be made. As I explain below, a process of differentiation and clustering utilizes these attributes to form architectures or structures to which meaning can be later appended.[17]

The first "batch" of sensorial inputs that are captured by the mind can be described as a first coat of paint on a wall or a first level or sediment of flooring. It is not yet a painted wall nor a usable floor—just the foundation. This batch of sensorial inputs, such as light and sound, have an initial quantity of, for example, luminescence or decibels of sound. However, these are "raw" quantities as there is a lack of benchmark for comparison.

The first batch of sensorial inputs forms in the mind a structure similar to the scale of musical notes. There is a very large number of permutations of such notes, as well as sensorial inputs, but only after a second batch is captured by the mind. Once such a second batch is received, the first batch can now serve as attributes, and the second batch is used to *refine* the attributes, like a musician testing an instrument to the exact notes.

In this manner, the "big bang of knowledge" produces the categories or foundational framework upon which subsequent sensorial signals can be manipulated and compared. For example, a sound at first is just a sound, whereas when compared to a foundational sound it is now an event which can be appraised as to volume, distance from the source, and other such attributes. The "quality" of the first batch, or the foundational batch of signals, is, of course, an essential element of the mind's ability to process later signals. But, what is such "quality" or characteristic of the formation of the foundation?

In a way, it is similar to the characteristics of a coat of paint—the quality of the paint used. In the human mind it would depend on the receptacle rather than on the paint—that is, it would depend on the quality of the wall as material that better absorbs and contains paint. Because human mental mechanisms and physical characteristics of the brain are genetically inherited and are similar across the species, we humans create similar foundational architectures of the first batch of signals to which we are exposed and which our mind thus absorbs. In the example above, the paint is the same, the walls are similar, the painters and their strokes (genetics) are different.

With the existence of the second batch, hence the ability to compare and to generate workable attributes, there is also the initial *learning* and *memory* events in the life of the mind. It is at this stage that the brain is also transformed into a mind—although knowledge is not yet engendered in the mind.

By memory I mean that the imprints of the first batch of signals are kept in the brain. This is the ability that the evolutionary process has endowed us: to be able to retain these biochemical imprints, thus to be able to link to them subsequent batches of sensorial inputs so that attributes can be applied to them.

The Process of Joining Attributes and Sensorial Inputs

The mind is exposed to sensorial inputs and their attributes. This is a complex influx of items. There is a very large number of permutations (almost infinite) of the sensorial inputs and their attributes. These permutations may be: (1) by a mix of inputs and attributes, or (2) by a mix of differences.

The mind, by the interface of biochemical exchange among neurons, collects and manipulates sensorial inputs and their attributes. The basic process by which such actions are reflected in the neurons is perhaps the same for sensorial inputs and their attributes.[18] The conjoining of sensorial inputs and attributes is possible when they exhibit *difference* or *change* from an initial state in which the mind was exposed to them. For example: (See Figure 2).

The differences in *each* sensorial input are specified in *intervals,* where the magnitude of the interval would play a significant role in the manipulation of sensorial inputs. For example, a unique input of a sound or a note in itself can only be joined with such an attribute as emanating from a given space or a point in space. When a second note is heard, it is now possible to join the differential in notes with such additional attributes as size or volume (high or low), location in space (near or far), and direction (getting closer or farther). There may be meaning to only one attribute attached to a sensorial impact, but the more intervals that can be derived, the more complex the sensorial inputs become and the more meaningful they are for further manipulation on the way to transforming these inputs into knowledge.

Figure 2.

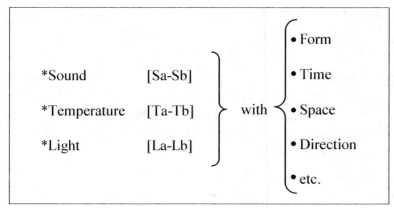

This is not a trivial point. The more sensorial inputs are received and the more varied they are (such as the case of the robot Michael), the more the mind is able to fine-tune the attributes and also to increase the number of permutations of sensorial inputs and the conjoining of these with attributes. Simply put—the more we absorb inputs from our senses, the more we lay the foundation for knowledge—in other terms, the more we create experiences that allow us to redefine and to refine our capturing of what constitutes the universe outside our mind. This is still a mechanical view—it does not mean that we gain more knowledge about the "true" world outside our mind or the real world. It simply means we get a more refined or focused view of the physical world that exists outside the mind and *from which signals are received* by the mind—namely, the *source* of the signals or inputs.

Intervals are differences between two states when an attribute is applied. The difference is not yet based on a *benchmark,* a pre-set criterion of structural or ethical significance. Rather, a sound heard by the human mechanism of the ear is placed or considered by the human mind in conjunction with another sound, where there is a difference in any of the attributes: loudness (size), location in space, time of appearance, distance, or direction. Therefore, the interval is simply a tool used by the intelligent being to *measure* the very basic element of knowledge.

In this model of the atomization of knowledge, two initial questions arise. First, what do the inputs and attributes describe (as they are absorbed in the form of intervals), and second, what are the criteria of mutual attraction of these inputs, or the rules by which they conjoin to form more complex modes of combinations or inputs and their attributes?

A basic unit of knowledge is thus a *perceptible distortion* in the set of inputs and attributes where intervals are perceptible (thus become significant) so that the vibrations in the senses become an item or unit of what can be considered a basic unit of knowledge. Multiple inputs and their attributes form complex views of the external world, which are ultimately expressed in higher-order terms such as human language (words and descriptions of concepts).

The mind recognizes sensorial inputs by means of the differences and increased complexity of the number of permutations. But the question persists: how such recognition occurs so that intervals in sensorial inputs plus attributes form some meaning—however primitive such meaning may be.

The only possible process that would apply to the conjoining of inputs and attributes is "*trial-and-error.*" This unique process allows the *perceptible distortions* to be stored as experience and to serve as benchmarks for future

exposures to sensorial inputs. The first such event is devoid of any comparative meaning. It forms the very initial formation of the conjoining of inputs
and attributes.[19] As later inputs and attributes are received and absorbed, they
may now be compared with the initial experience. Thus, the differences and
similarities with the previous sensorial intervals (SIs) reveal characteristics
of these SIs that permit the brain to now cluster them in a more meaningful
manner.

The process of "trial-and-error" operates on the principle of a stepwise
comparison with previous inputs and their attributes. In the absence of an
a priori set of categories, experiences are added cumulatively in a way that
the differences or intervals are viewed by their similarities and differences,
and such comparisons refine the mind's perspective of what constitutes the
external experience described by the sensorial inputs and their attributes.
The more these experiences accumulate, the more they are disposed to be
congregated or clustered into more meaningful constructs.

The state of science in brain research does not yet allow for a definitive description of the biological or biochemical transactions responsible for such
congregation into higher-order constructs. We do know that there are interfaces between neurons, in which sodium ions are transferred from synaptic
transmitters to other neuron receptors. This may be a mechanism by which
neurons "fire" and interact with each other during mental activities, such as
learning or gaining knowledge. Scans with functional MRI (magnetic resonance imaging) can track these transfers of sodium ions, thus mapping neural
activity and matching it to cognitive changes (Thulborn, Adams, Grindin, &
Zhou, 1999; Little, Klein et al., 2004; Little & Thulborn, 2005).

These biological processes may illuminate the manner in which sensorial
inputs and their attributes congregate. Exchanges of chemical elements among
neurons may be basic mechanisms by which sensorial inputs are registered
and classified by their attributes. These mechanisms may also begin to explain
the attraction principles outlined below. Perhaps chemical attractions are the
elemental principle, and cognitive attraction follows.

Congregation or Attraction

Sensorial intervals conjoin or congregate in a mode that is perhaps similar
to that of sub-atomic particles forming atoms. But, what brings these SIs

together? Are there laws that govern or influence their attraction or rejection (such as the law of gravity)?

As SIs accumulate and are compared with previous SIs, similarities in *shared attributes* may emerge. The sharing may occur in such attributes as *space, time, speed,* and *form.* Certain characteristics may now be applied to the sharing of attributes. For example, the shared attribute in space may be viewed as "*proximity.*" A similar characterization would apply to SIs that are close in *time.* The magnitude of permutations of such sharing is enormous, allowing for a very large number of possibilities.[20]

Although this proposed process of cumulation may originally be preceded by chemical attraction and exchange between neurons, it is nevertheless a different mechanism, both in level of aggregation and in its functioning. This principle of attraction by, for example, shared attributes may perhaps occur at the cellular or molecular levels, and involve a large number of individual neurons (as carriers of sensorial inputs and their attributes), arranged and then interacting in a massive array of permutations. The architecture of these permutations will provide a rationale and order that will engender higher-order constructs.

Representation, Limits, and Hierarchies

Units of knowledge may have several attributes by which the mind is able to congregate them. Each congregation of units forms a higher-order construct that creates a unique representation of a phenomenon in nature—hence creating knowledge about such a phenomenon. Different congregations or combinations of basic units create different perspectives of the same phenomenon.

There are limits to the degree of representation of natural phenomena—hence to knowledge generated by the basic units and their combinations. Such limits are limitations of the senses, the affinity of basic units of knowledge to be attracted to each other and to congregate to form higher-level structures, and the ability of the mind to aggregate them.

A question arises: Is the attraction between basic units of knowledge ontological, or is it inherent in the attributes? Also, are there *portals of connectivity* for attraction? (Gibbons et al., 1994).

If there are *portals,* what are their characteristics and what are the rules or principles that govern the flow through them? Are these portals similar—in form or function—to benchmarks?

The possible existence of principles of attraction suggests that the basic elements or building blocks of knowledge can provide a representation of reality (nature) within the limits of our senses—and within the limits of attraction or conjoining of the basic elements. Is it also possible that attraction occurs by different attributes, simultaneously? For example, when basic elements of knowledge (sensorial intervals) vary by time, format (size), distance, or direction—simultaneously—will such multiplicity of variability create a stronger or more powerful attraction? In other words, does it make a difference, for the formation of higher-order and more complex structures in knowledge, whether attraction is *mono-attribute* or *multi-attribute*?

This question is related to the issue of *richness* of the basic unit of knowledge. How would multiple sensorial intervals add to the richness of the basic unit of knowledge? For example, if a basic unit is composed only of intervals of light vs. multiple intervals of a variety of signals such as *color, temperature,* and *sound,* will the latter scenario offer a richer or more meaningful representation of reality, and would such richness impact the ability or propensity of these basic units to congregate and combine to form higher-order structures? Also, will the mind be more apt to make more sense of complex or rich aggregations?

Perhaps there is a *hierarchy* of the representation of reality by the elemental units of knowledge. A single sensorial interval will combine with other sensorial intervals to form, as a first step in a *hierarchy of complexity,* a richer and perhaps more viable representation of nature or reality. In this case, what is the principle by which such multiple elements combine? Perhaps they do conjoin in the *knower,* where these multiple units are related (in the mind of the knower, or the processor of these units) to a single natural phenomenon, occurrence, or object of inquiry on the part of the knower.

A second step in such a hierarchy is the congregation of these conjoined, multiple sensorial intervals. But, what is the nature of the structure formed by the aggregation of multiple sensorial intervals—all linked by a principle of a common natural phenomenon—as it is justified in the mind of a knower or processor of these units? There is a need to establish the framework of such a conjoined structure. We may refer to this complex unit as KANE (*K*nowledge b*A*sic u*N*it of *E*xistence).

The Emergence of a Basic Unit of Knowledge

Attraction of different sensorial intervals occurs to form a KANE, which is the first structure that allows for a meaningful glimpse of knowledge. Why would intervals in sound and light, for example, be attracted to each other? There are no ontological similarities or biological/chemical exchanges that would cause these sensorial intervals to conjoin in a mode similar to that of attraction of sub-atomic particles. Therefore, the conclusion must be that KANEs are formed in the knower, where certain sensorial intervals (SIs) are considered related, hence they are conjoined. Certainly, this requires a process of some logical or intelligent analysis of SIs and a given set of rules by which selected SIs are conjoined to form KANES, while other SIs are not combined—and why and which SIs are combined with which SIs.

It is conceivable that there are *two* modes of attraction. The first is attraction and congregation of SIs within their attributes. This is a structural issue that requires further examination, as well as a mathematical elaboration.

This first mode of attraction is the conjoining *within* the attributes. As such it is more intuitively understood. SIs are attracted to each other because of similarities of the context attributes such as time, space, speed, direction, and distance of signal. There is a *commonality* in the SIs. The knower who receives SIs with shared or common attributes will tend to conjoin them.

The second mode of attraction and congregation of SIs is within an external set of criteria of attraction (external to the SIs). As in the case of attraction because of attributes of SIs, an external criterion may consider some common parameters shared by the SIs, but that are imported by the knower or processor of the signals. For example, a KANE may be formed because SIs are conjoined because of their relation to an event or phenomenon such as a similar KANE that exists in the memory of the knower (Kant, 1999).

For instance, a sensorial interval may also vary in space and be attracted to another sensorial interval *also varied in space.* Similarly, the first SI (varied in space) may be attracted to a SI varied in time. Are there different rules in such seemingly distinct attractions that govern the attraction (as to its power or the resultant KANE)? By power I refer to the strength with which the KANE describes reality. Is such power of representation or description related to the level of attraction? Is attraction, in the sense described above, a different term for *relatedness*? This definition may be important in further use of mathematical/statistical tests of relationship.

In another perspective, the attraction is based on the chemical properties of the material being exchanged by neurons, and on the biological make-up of transmitters and receptors within neurons. The questions still remain: How are KANEs formed? Why do certain neurons "fire" at other neurons? And what are the contents of the sensorial inputs that are being exchanged?

Criteria for Attraction

The analysis presented above addressed two modes of attraction within attributes and due to external factors. These modes also imply two distinct types of criteria for attraction and congregation. If we distinguish between the notions of *attraction* and *conjoining,* then the criteria for these distinct notions will also be different.[21]

In the model of the atomic nucleus, for example, the sub-atomic components are attracted to each other by virtue of their *different* electric charges. This is a physical attraction. In the case of the sensorial inputs, the attraction is not physical but a relationship engineered by the mind, or the knower.

How does this happen? Sensorial inputs are merely imprints on the neural framework of the mind. This mind has the ability to discern the similarities in the inputs. Similarities and differences are noted. The mind is able to discern whether there is an attraction or whether the sensorial inputs are disconnected, dissimilar, or not attracted to each other. This is simply the step in which the mind establishes whether there is any attraction between sensorial inputs—by virtue of their attributes.

At this point there is no meaning attached to the sensorial inputs. Conjoining or clustering of the sensorial inputs will be the next step, following the establishment of attraction. Thus, inputs on *context attributes* (such as speed, time, space, or distance) will be used to establish attraction, whereas *relation attributes* (such as form, intensity, or life of the signal) will be used to initiate clustering or conjoining.

As I have stated, sensorial inputs are mere imprints on the neural system of the knower who has the ability to instill some order into these signals. The knower is able to establish attraction and to initiate clustering—thus it begins to assign sense or meaning to the flow of sensorial inputs.[22] In effect, sensorial inputs describe events that occur outside the mind (within the knower or in the world outside). Making connections between and among events is the ability of the mind to discern attraction and to initiate conjoining or clustering. Without such ability, the sensorial inputs will just be inputs, imprinted upon the mind and remaining as imprints, hence lacking the capability of becoming knowledge.[23]

Heredity or Heuristics

How does the mind discern attraction in the attributes of sensorial inputs? How does the mind initiate conjoining or clustering of these sensorial inputs? In previous pages I suggested that cognition is initiated by "trial and error" and that there is a phenomenon of the "Big Bang" by which mechanistic imprints of SIs are clustered for the first time.

The mind does not possess inherited constructs or concepts that allow it to compare SIs with these a priori assets. Rather, I suggest that the ability to discern attraction in the attributes of SI (hence to cluster them) is derived from the same mechanistic characteristics that allow the mind to receive and discern sensorial inputs. The ability to cluster and the formation of the first instance of knowledge creation are discussed in Chapter VI.

Thus I conclude here with the suggestion that heredity is active in the mechanical or biological ability of the mind to discern and to cluster SIs. The structure and interface of neurons firing chemical exchanges are such that they allow for such capabilities to discern and to cluster SIs.

The *ability* to affect attraction and clustering is inherent in the physical structure and biological and biochemical processes of the mind. These are the result of heredity. But, genetics provides only the means and the equipment, whereas their use in an effective manner is an individualized effort.

It seems that either genetic inheritance or later deficiencies in the working of the mind would hinder the processing of knowledge. The fact that the human mind is able to process knowledge beyond the basics needed for survival of the species may be an aberration or an extraordinary event in the evolution of our species. Genes carry in them inherent traits which may constitute an elemental form of knowledge. Yet, the working of the mind is a much more elaborate process of attraction and clustering which may or may not have a function in the survival of the species. Being able to conjure higher-order constructs is not a necessary condition for survival or procreation of the species. In humans, evolution has favored such ability.[24]

References

Alston, W. (1989). *Epistemic justification.* Ithaca, NY: Cornell University Press.

Audi, R. (2002). *The architecture of reason: The structure and substance of rationality.* New York: Oxford University Press.

Averill, E. (2005). Toward a projectionist account of color. *Journal of Philosophy, 102*(5), 217-234.

Barsalou, L. (1999). Perceptual symbol systems. *Behavioral and Brain Sciences, 22*(3), 577-660.

Barsalou, L. (2003a). Abstraction in perceptual symbol systems. *Philosophical Transactions of the Royal Society, 358,* 1177-1187.

Barsalou, L. (2003b). Situated simulation in the human conceptual system. *Language and Cognitive Processes, 18*(5/6), 513-562.

Bell, D., & Vossenkuhl, W. (Eds.). (1993). *Science and subjectivity.* Berlin: Akademie Verlag.

Blachowicz, J. (1998). *Of two minds: The nature of inquiry.* Albany: State University of New York Press.

Buzaglo, M. (2002). *The logic of concept expansion.* New York: Cambridge University Press.

Churchland, P. (1997). Knowing qualia: A reply to Jackson. In N. Block, O. Flanagan, & G. Guzeldere (Eds.), *The nature of consciousness: Philosophical debates.* Cambridge, MA: MIT Press.

Cooper, D., Mohanty, J., & Sosa, E. (Eds.). (1999). *Epistemology: The classic readings.* London: Blackwell.

Dennett, D. (1988). Quinning qualia. In A. Marcel & E. Bisiach (Eds.), *Consciousness in contemporary science.* New York: Oxford University Press.

Dennett, D. (1991). *Consciousness explained.* New York: Little, Brown, and Company.

Gertler, B. (1999). A defense of the knowledge argument. *Philosophical Studies, 93*(2), 317-336.

Gibbons, M., Limoges, C., Nowotny, H., Schwartzman, S., Scott, P., & Trow, M. (1994). *The new production of knowledge: The dynamics of science and research in contemporary societies.* London: Sage.

Harnad, S. (2000). Correlation versus causality: How/why the mind/body problem is hard. *Journal of Consciousness Studies, 7*(4), 54-61.

Heath, T., Melichar, J., Nutt, D., & Donaldson, L. (2006). Human taste thresholds are modulated by serotonin and noradrenalin. *Journal of Neuroscience, 26*(49), 12664-12671.

Humphrey, N. (2000). How to solve the mind-body problem. *Journal of Consciousness Studies, 7*(4), 5-20.

Jabs, A. (1992). An interpretation of the formalism of quantum mechanics in terms of epistemological realism. *British Journal for the Philosophy of Science, 43*(3), 405-421.

Jackson, F. (1982). Epiphenomenal qualia. *The Philosophical Quarterly, 32*(2), 127-136.

Kandel, E. (2006). *In search of memory: The emergence of a new science of mind.* New York: W.W. Norton.

Kant, E. (1999). *Critique of pure reason.* Cambridge, UK: Cambridge University Press.

Lee, K., Kageura, K., & Choi, K. (2004). Implicit ambiguity resolution using incremental clustering in cross-language information retrieval. *Information Processing & Management, 40*(1), 145-159.

Little, D., Klein, R., et al. (2004). Changing patterns of brain activation during category learning revealed by functional MRI. *Brain Research: Cognitive Brain Research, 22*(1), 84-93.

Little, D., & Thulborn, K. (2005). Correlations of cortical activation and behavior during the application of newly learned categories. *Brain Research: Cognitive Brain Research, 25*(1), 33-47.

Marks, L. (1978). *The unity of the senses.* London: Academic Press.

Nida-Rumelin, M. (1996). Pseudonormal vision: An actual case of qualia inversion? *Philosophical Studies, 82*(2), 145-157.

O'Reagan, K., & Noe, A. (2001). A sensorimotor account of vision and visual consciousness. *Behavioral and Brain Sciences, 24*(5), 979-980.

Ohm, E., & Thompson, V. (2004). Everyday reasoning with inducements and advice. *Thinking & Reasoning, 10*(3), 241-272.

Oliver, J. (1960). The problem of epistemology. *Journal of Philosophy, 57,* 297-304.

Pinker, S. (1994). *The language instincts: How the mind creates language.* New York: William Morrow.

Polanyi, M. (1966). *The tacit dimension.* New York: Anchor Day Books.

Quine, W. (1977). *Ontological relativity.* New York: Columbia University Press.

Reid, T. (1941). *Essays on the intellectual powers of man.* London: McMillan.

Rose, S. (Ed.). (1999). *From brains to consciousness? Essays on the new sciences of the mind.* Princeton, NJ: Princeton University Press.

Schunn, C., & Vera, A. (2004). Cross-cultural similarities in category structure. *Thinking & Reasoning, 10*(3), 273-287.

Small, G. et al. (2006). PET of brain amaloid and tau in mild cognitive impairment. *New England Journal of Medicine, 355*(25), 2652-2663.

Squire, L.R. (1987). *Memory and brain.* New York: Oxford University Press.

Thulborn, K., Adams, D., Grindin, T., & Zhou, J. (1999). Quantitative tissue sodium concentration mapping of the growth of focal cerebral tumors with sodium magnetic resonance imaging. *Magnetic Resonance in Medicine, 41*(2), 351-359.

Whitehead, A. (1979). *Process and reality* (2nd ed.). New York: The Free Press.

Endnotes

[1] The separation of the search for the basic unit of knowledge from larger problems of philosophy is not simply an intrusion by an organization scientist into another disciplinary domain. Nor does such a phenomenon imply any attempt or intent to deprecate or underestimate the value of philosophical inquiry. Rather, separation is a methodological effort which may or may not result in theoretical advances that would influence the inquiry into larger philosophical issues. See, for example: Walton, D. (2004). A new dialectical theory of exploration. *Philosophical Explorations, 7*(1), 71-89; Guyer, P. (1987). *Kant and the claims of knowledge.* Cambridge, UK: Cambridge University Press; Oliver (1960); Jabs (1992); Whitehead (1979).

2   At this stage issues such as theories of the mind and the difference between pictorial and linguistic representations of sensorial inputs of the external world are premature and will be discussed when I explore the clustering of sensorial inputs. See, for example, O'Reagan, J., & Noe, A. (2001). A sensorimotor account of vision and visual consciousness. *Behavioral and Brain Sciences, 24*(5), 939-973; Banerjee, A. (2002). The roles played by external input and synaptic modulations in the dynamics of neuronal systems. *Behavioral and Brain Sciences, 24*(5), 811-812; Ryle, G. (1990). *The concept of the mind.* London: Penguin Books; von Melchner, L., Pallas, S., & Sur, M. (2000). Visual behavior mediated by retinal projections directed to the auditory pathway. *Nature, 404,* 871-876. In addition, at this point the philosophical issues of raw feelings or "qualia" and the nature and operational role of consciousness are also deferred to the later examination of clustering of sensorial inputs.

3   In a crude way, it is possible to assert that "to see is not to know; to hear is not to know; to see and hear is not to know—unless one has a clue as to what one sees and hears." However, I am not arguing that this first stage (in which signals are received and acknowledged as such by the senses) is merely *physical* manifestations of inputs, purely as exchange of chemicals, rather, by the acknowledging (albeit physically/chemically), the stage is set for clustering and assignment of meaning to sensorial inputs.

4   Werkmeister (1948) argued that sensorial inputs are descriptors of our (my) experience of "things" in the world about us (me). But these sensorial inputs by themselves are not sufficient to ascertain the existence of the "things" they describe. This is the problem of the qualia, or the intrinsic properties of experience (O'Reagan & Noe, 2001).

5   It is difficult to ascertain where the model I propose here of the structure of knowledge coincides with Kant's philosophical framework and where it departs from it to form new perspectives and to break new ground. The immense literature that flourished about Kant's method and philosophical framework, including his philosophy of knowledge, shed light on the many nuances of his philosophy. I believe that the model of structure described here is a continuation of Kant's work, albeit engendered from a different discipline, with different research queries and in a different time.

6   See, for example, Heath, Melichar, Nutt, and Donaldson (2006), who describe the biochemical influences on changes in the sense of taste. Also see Small et al.'s (2006) study of cognitive impairment.

7   This is the issue of reductionism to a certain level of microanalysis. What are these frames of reference and are they composed of even smaller ingredients? Some answers are given in the discussion on the "Big Bang" of knowledge. Also, the reader who wishes to further explore this topic of perception, cognition, and the processing of signals by the mind should consult, for example: Hommel, B. et al. (2001). The theory of event coding (TEC): A framework for perception and action planning. *Behavioral and Brain Sciences, 24*(5), 849-937; Barsalou, L. (1999). Perceptual symbol systems. *Behavioral and Brain Sciences, 22*(3), 577-660. Also see: Lamberts, K., & Shanks, D. (1997). *Knowledge, concepts, and categories.* Cambridge, MA: MIT Press; Martin, A. et al. (2000). Category specificity and the brain: The sensory-motor model of semantic representations of objects. In M. Gazzaniga (Ed.), *The new cognitive neuroscience* (2nd ed., pp. 1023-1036). Cambridge, MA: MIT Press; Wisniewski, E. (1998). Property instantiation in conceptual combination. *Memory and Cognition, 26*(2), 1330-1347.

8   Barsalou and others describe this traditional approach (based on perceptions) as *modal* and *analogical,* whereas the contemporary nonperceptual approach is *amodal.* I will refrain from using these and similar esoteric terms in my description of the model I advance in this chapter, and its relation to the contemporary theories of cognition.

9   The reader interested in additional reading should consult: Hochberg, J. (Ed.). (1998). *Perception and cognition at century's end: Handbook of perception and cognition.* San Diego: Academic Press; Fodor, J. (1998). *Concepts: Where cognitive science went wrong.* New York: Oxford University Press; Margolis, E., & Laurence, S. (1999). *Concepts: Core readings.* Cambridge, MA: MIT Press.

10   I will return to this distinction in my discussion of "contextual" vs. "content" criteria for clustering. The distinction between "conscious vs. unconscious" categorization and my classification is that the former relates to the *cognitive mechanism* that is processing the inputs, whereas my classification refers to the *inputs* themselves.

11   The interested reader may consult the following sources: Barsalou (2003); Milner, A., & Goodale, M. (1995). *The visual brain in action.* Oxford, UK: Oxford University Press; Laming, D. (1986). *Sensory*

analysis. London: Academic Press; Kosslyn, S. (1994). *Image and brain.* Cambridge, MA: MIT Press. Barsalou (1999) had commented that: "Specifying how the cognitive system knows where to focus attention during the symbol formation process constitutes an undeveloped component of the theory...that remains far from resolved in any theory" (p. 608, note 7).

[12]   These problems have been topics of much discussion by both philosophers and, more recently, cognitive psychologists. The reader should consult Kant (Cambridge University Press, 1998 edition), Cooper, Mohanty, and Sosa (1999), Audi (2000), and Blachowicz (1998).

[13]   At this juncture of the narrative, I am not concerned with such issues as the epistemic truth, authority, or cogency of the arrangement or structural configuration that is formed by the initial manipulation of sensorial inputs by means of the attributes. The reader may wish to consult Schunn and Vera (2004), Ohm and Thompson (2004), and Alston (1989).

[14]   The terms "brain" and "mind" are used interchangeably in this chapter. I fully recognize and appreciate the distinctions between the physical entity ("brain") and the more conceptual construct of the "mind." The two are used here in the sense of an entity that is capable of collecting sensorial inputs, then also manipulating, storing, and recalling them as they were arranged and manipulated.

[15]   As I will confess throughout this book, I do not profess to have all the answers. The model I propose here—in my view—advances the state of our understanding of what constitutes knowledge and how it progresses by examining the stepwise structuring and build-up of knowledge from sensorial inputs to higher-order concepts. In several of these stages, I consider my proposal to be the logical possibility in the way knowledge is formed.

[16]   For such circular reasoning, the reader may consult Werkmeister (1948), Audi (2000), Buzaglo (2002), and Bell and Vossenkuhl (1993).

[17]   There is a host of problems that arise from my assertion that attributes reside in the environment being scanned by the sensors. I recognize and appreciate such philosophical issues, but not as impediments to the coherence of the model presented here. Differentiation and clustering will provide adequate answers to many of these valid problems (e.g., reliability, truth, formation of concepts, and verification).

[18] Research into neuron activity and biochemistry of the brain at the cellular level should be thus directed toward discovering those exchanges of biochemical compounds that allow the brain to recognize and to conjoin inputs and their attributes.

[19] There is an issue of how we would reconcile between "first experiences" with a small number of attributes vs. more complex (multi-attribute) inflows that now would be compared with the initial experience. Notwithstanding issues of verification, trust, and true reflection of reality, a possible solution would be to have more complex inflow replace less complex former experiences as the "benchmark experience" for the type of inflow.

[20] It would be a challenging task to address the mathematical implications of these arrangements of SIs as the fundamental units of knowledge. At the time of the preparation of this book, I was working with a colleague in developing potentially useful mathematical models to the interaction of signals and attributes, and the laws of attraction that govern such interactions.

[21] In the remainder of this chapter and in Chapter V, I will refer to the notion of conjoining (or aggregation) as "clustering." My choice of this specific term is not ontological nor methodological. It seems to me that the term better describes the transformation of distinct items into a more complex and meaningful entity. The dictionary defines "cluster" as: "a number of similar things growing together or of things or persons collected or grouped closely together" (Webster's New Collegiate Dictionary, 1977).

[22] The neurobiology and neurobiochemistry of these interfaces need to be explored. Currently we believe that exchanges of certain compounds among neurons (proteins and amino acids) may be responsible for the ability to discern and to cluster individual imprints of sensorial inputs. By suggesting an overall framework and the process by which knowledge is structured, I am perhaps setting some form of a roadmap for neural research. Thus, perhaps attraction is discerned by the exchange of certain compounds among selected neurons, whereas clustering will be induced by the exchange of a different set of compounds among other neurons.

[23] I must distinguish between this state of SIs remaining imprints and the often-used typology of data-information-knowledge. Imprints of SIs are not necessarily data or information. They are simply SIs imprinted

upon the mind (or knower) as potential raw material for the structuring of knowledge. Thus, machines or other biological beings may receive such sensorial inputs, imprint them upon their version of a mind, and not act upon them for lack of the said ability.

[24]   This analysis would eventually lead to the development of a typology and taxonomy of knowledge. In this case it would be at least a classification into knowledge which is necessary for survival (such as genetic characteristics), and knowledge which is initially superfluous but ultimately results in better skills for survival (such as higher-order constructs and a learned understanding of the world and how to control it). These conclusions will be incorporated into Part IV of this book. See, for example, Kandel (2006) and Squire (1987).

Chapter V

Synaesthesia:
The Gateway to Knowledge

What Is Synaesthesia?

A good portion of brain research has been directed towards its anomalies, and mental illnesses that result from them. Examples include Alzheimer's disease and synaesthia.[1] This is due partly to the desire of researchers to assist in understanding diseases and hastening cures for them, and partly because the study of anomalies is a better way to understand how complex systems such as the human brain actually function (Harrison, 2001).

Synaesthesia means the union of sensations or the joining together of human sensations such as sight, taste, or hearing. This phenomenon is considered an abnormal functioning of the brain. It seems to affect, in some form, about 3% of the population, and in the more advanced form, about one in 200 people (Harrison, 2001; Ward, 2004). This is a condition whereby when one sense is stimulated (such as hearing), it will also engender stimulation in another or other senses. People who have this anomaly can, for example, associate the sensation of taste upon hearing a certain sound or associate color with hearing a sound. Simner and Ward (2006) described such a phenomenon. They studied six individuals who associated the sounding of words with taste sensations, monitoring the association via brain imaging.[2]

Synaesthetes (people who have this anomalous condition) make such associations of senses involuntarily and in various combinations of the senses. They are exceptional individuals in this ability to "cross-over" among the senses so that they perceive sensations without the appropriate sensorial inputs from their external environment.[3] This condition occurs in families, thus it is inherited, perhaps the result of multiple genetic abnormalities. In addition, this anomaly has been associated with other cognitive deficiencies, such as dyslexia.[4]

Synaesthesia and Knowledge

Although synaesthesia has been considered an abnormal condition, it holds some cues to the manner in which knowledge is generated in the human mind. The clustering of sensorial inputs provides the initial formation of what we consider to be knowledge. We know because we are able to bring together different and *independent* sensorial inputs. The biochemical processes by which neurons fire and interact make it possible for such clustering to occur.

In synaesthesia, there is a clustering or conjoining of different sensorial inputs without the objective of necessarily creating constructs of knowledge. Although research has shown that synaesthetes are frequently also endowed with outstanding positive skills (such as perfect musical pitch and extraordinary memory abilities of instant recall), there does not seem to be a functional connection between this condition and the creation of knowledge.

But, when different sensorial inputs are conjoined, there is an inevitable creation of the initial structure of knowledge. Synaesthetes seem to respond with sensorial conjoining to specific words. Ward (2004) reported the case of a synaesthete who reacted to emotionally laden words such as love with "seeing" certain colors. In such a case the evoking of words is connected to pre-existing knowledge of the meaning of such emotions. Therefore, the response in terms of sensorial inputs is predicated on prior knowledge and meaning. Nevertheless, there is the phenomenon of the conjoining of diverse sensorial inputs, when only one of them is actually experienced by the person.

It is thus conceivable that this anomalous phenomenon is the manner in which human knowledge actually began. Different senses conjoined via a mechanism that was perhaps triggered by a genetic mutation. Instead of just one sense providing inputs to the brain, another or perhaps multiple other sensorial inputs conjoin—without the actual experience—to form a cluster, hence a KANE. All this perhaps was accomplished automatically, due to the genetic anomaly. Evolutionary processes may have standardized the process by which the clustering of different sensorial inputs becomes a highly beneficial event, since it engenders knowledge, which in turn helps the species to survive. However, the original event itself, like the disappearance of the rear tail in humans, was relegated to a rare phenomenon, but still recessively active in genetic exchange.

Knowledge *is* clustering of sensorial inputs. Human knowledge could not have emerged without experiences from the world outside the mind. The firing of neurons with inputs from senses that have not been energized with external experience could only happen when *previously* there had been such prior experience.[6]

There remains the question of how such sensorial inputs are retained in memory and what the biochemical processes are that make such retention and then revisiting or extraction of sensorial inputs. Whether sensorial inputs are accessed from memory or from actual experience is not an issue in the view of knowledge or clustering of sensorial inputs. In this book I argue that knowledge is engendered when sensorial inputs are conjoined. The sources of such inputs are the contemporary experiences *and* the inputs retained in memory.

A related research problem would be the different formats, functions, effectiveness, and values of knowledge that are created solely and entirely with sensorial inputs from contemporary experience (from the world outside the mind), and knowledge created by either a mix with memory or from memory alone. It is also conceivable that abnormal perceptions, imagination, illusions, and dreams would create "pseudo" knowledge based on the conjoining or clustering of sensorial inputs embedded in memory, or as a combination of actual/contemporary experience mixed with memory.[7]

References

Harrison, J. (2001). *Synaesthesia: The strangest thing.* New York: Oxford University Press.

Luria, A. (1968). *The mind of a mnemonist.* New York: Basic Books.

Raines, T. (1909). Report of a case of psychochomesthesia. *Journal of Abnormal Psychology, 4*(2), 249-260.

Ramachandran, V., & Hubbard, E. (2001). Synaesthesia—A window into perception, thought, and language. *Journal of Consciousness Studies, 8*(12), 3-34.

Sagiv, N., Simner, J. et al. (2006). What is the relationship between synaesthesia and visuo-spatial number forms? *Cognition, 101*(1), 114-128.

Simner, J., & Hubbard, E. (2006). Variants of synaesthesia interact in cognitive tasks: Evidence for implicit associations and late connectivity in cross-talk theories. *Neuroscience, 143*(3), 805-814.

Ward, J. (2004). Emotionally mediated synaesthesia. *Cognitive Neuropsychology, 21*(7), 761-772.

Whipple, G. (1900). Two cases of synaesthesia. *American Journal of Psychology, 11*(4), 377-404.

Endnotes

[1]   The American Synaesthesia Association is a professional organization composed primarily of brain and cognitive researchers. The emphasis of their research is on this phenomenon as a neurological condition, treated as an anomalous sensory perception. The association is not concerned with the implications of synaesthesia on knowledge and knowledge creation. Ramachandran and Hubbard (2001), for example, had suggested that synaesthesia is a phenomenon in human perception, "…not an effect based on memory associations from childhood or a vague metaphorical speech" (p. 3). They also suggested that synaesthesia is caused by a "…genetic mutation that causes defective pruning of connections between brain maps" (p. 3).

[2]   Also see Simner and Hubbard (2006) and Ward (2004).

[3]   Among famous synaesthetes are the painter Vladimir Kandinsky, novelist Vladimir Nabukov, and the French poets Arthur Rimbaud (1854-1891) and Charles Baudelaire (1821-1867). Another case is described in Luria (1968). This instance is cited later in the book.

[4]   Synaesthesia has also been associated with positive cognitive functions such as musical proficiency. See, for example, Sagiv, Simner et al. (2006). Before researchers were able to document the stimulation in the brain of multiple senses by contemporary imaging technology, some cognitive researchers believed such a phenomenon to be an illusion of the mind.

[5]   See, for example, Cytowic (2002) and the earlier reports of this condition in Whipple (1900) and Raines (1909).

[6]   After, of course, the "big bang" of knowledge.

[7]   This research problem is conducive to the exploration of existing research in cognition and neuropsychology, as well as brain research—as a cross-disciplinary effort with behavioral sciences and knowledge management. The framework advanced in this book may be helpful in the framing of this research direction and the specific research challenges it offers.

Chapter VI

Knowledge is Clustering

Mechanics and Reason: On the Road to Coexistence

The emergence of the basic unit of knowledge (by conjoining multiple sensorial intervals) is possible by means of the mind's ability to cluster such sensorial intervals. As I outlined in Chapter IV, this is a mechanistic or structuralist approach.[1]

The mind has the ability to cluster sensorial inputs and to form reasoned constructs of increasing complexity. Again I raise the question: What is the road from the mechanics of clustering to the articulation of reason? The road is a process by which clustering of signals and their attributes is conducted. Although the neurochemical processes are still largely unknown, I advance a framework that may be a poor substitute, yet adequately explain, the mode in which clustering generates reason and knowledge (see Clarke & O'Malley, 1968; Shafto, 1986; Lynch, MacGaugh, & Weinberger, 1984).

Neuroscientists have actively investigated the structure of the brain and the locus of different functions performed by this organ. Churchland (1988) described the cerebral anatomy and had attempted to link activities in special areas of the human brain to higher functions of this organ. In this regard, the aim of her analysis was the "association cortex," which are those areas in between such regions of the brain as the somatosensory cortex and the visual cortex (Churchland, ch. 5, pp. 173-208).[2]

While considering the issue of the two hemispheres of the human brain, Churchland suggested that "ample evidence has accumulated that even in the normal case there is a great deal of complex, dense, *cognitive* processing that is not available to awareness, such as the processing that underlies the production and comprehension of grammatical sentences" (p. 181). Further, she also addressed the mind-body problem by exploring the issue of reduction of mental states to neurobiological states. This is the other direction of my earlier question: What is the road from mechanics to articulation of reason? Churchland thus disagreed with the common notion that "mental states are not physical states, either because they are the states of a nonphysical substance or because they are emergent nonphysical states of the brain in the sense that they cannot be explained in terms of neuronal states and processes" (p. 34). Rather, she concluded that although she believes that mental states may be functional states, it does not necessarily imply that "psychology cannot be reduced to neuroscience" and that "new theories about the nature of information processing and about the nature of information-bearing structures are badly needed" (p. 399).

These conclusions bring us back to the initial queries of mechanics and reason, and the path in which initial manipulations of sensorial inputs lead to the generation of higher-order rational constructs. This also reintroduces some philosophical questions of the place of consciousness and the gap, or perhaps harmony, between what Kant called "things in themselves"—the real world—and the mind.[3]

Humphrey (1999) addressed this topic of mind-body relationship in his interesting book on the history of the mind. He correctly observed that much depends on our definition of what constitutes "consciousness" and how we define higher-order constructs of this type. In the path from sensorial inputs to rational constructs, we are dependent upon the definitions of such constructs, and we may find biases that are anchored in archaic uses, cultural attributes, and other such variables.

Humphrey wondered, as I do in this book: "What is going on when we form the representations that *are* conscious? How is this representation done, where does it take place, how long does it last, and so on?" (p. 130). He argued that "every distinguishable sensation in human beings does correspond to a physically different form of sentiment" (p. 179). His hypothesis in linking sensorial inputs to sentiments or conscious bodily reactions is based on the different modalities of sensorial inputs. Simply stated, Humphrey hypothesized that sensorial inputs are a personal experience that translates into perceptions and a conscious recognition of such sensations, and that the mind forms such recognition by the link between the sensation and the sentiment it generates. He illustrated this by the sentiment of "alarm" generated by sensorial input of "red" through the retina of the eye. When the body sees red, the mind senses danger.[4]

Humphrey had also suggested that the link he hypothesized (between sensations and sentiments) is perhaps an initial step toward pointing the way to objective phenomenology. In the following pages I have attempted to go a step beyond the link, and to advance a process of clustering as the structure by which sensations become fundamental units of knowledge. It is by no means a leap forward to objective phenomenology, but I believe it to be a small step toward explaining the formation of the basic units of knowledge.

The "road to reason" has also been addressed by a host of other scholars. It is interesting to note how Bergson (1998) in the early years of the twentieth century distinguished between certain notions and the way we recognize them in our mind. In his discourse on 'intellect' and 'materiality', Bergson argued that our intellect as well as sensorial inputs are an instantaneous picture of reality that the human consciousness selectively retains. This process is therefore highly subjective, not allowing us to 'know' reality but simply to photograph it—hence the "illusion of our understanding," as Bergson had coined this phenomenon.[5]

A different approach to the explanation of the "road to reason" was initiated by biologists and neuroscientists and received a boost in the past two decades.[6] In the area of vision, there have been many promising developments. O'Reagan and Noe (2001) studied the relation between sensorimotor inputs and what they described as "visual consciousness." They advanced the notion that the ability to see is based on a set of skills the brain utilizes to process the visual inputs.[7] In their model there are two types of visual consciousness: *transitive* and *general.* The transitive consciousness is the ability to distinguish or be aware of one portion or aspect of the outside world being scanned by the eyes. The general consciousness is the awareness of more than one portion or aspect of the scene being scanned.[8]

This model explains the brain's capacity to capture the elements of a picture in the outside world in terms of its ability to match these visual sensorial inputs with the practical knowledge we have of how to process these particular inputs (sensorimotor contingencies). This practical knowledge is based on experience (of prior events) in this particular type of encounter with the external world. We therefore comprehend what we see in the raw sensorial inputs as the scene in the outside world because we are able to channel these inputs through the existing framework of knowledge that tells us what to do with these inputs. If I understand their model, the way the mind formulates notions of things that we see (a tree, the color red) is by applying a selected set of rules of how to process these inputs, like a diamond cutter using tools and an approach to cutting *this specific* raw stone.

In a way, this model sets the tone for a research program that would explore the pathways and the mechanisms by which the brain processes sensorial inputs. The different skills and their disposition for different types of sensorial inputs would correspond to exchanges or flow of different proteins.

My search for the elemental structure of knowledge has its parallel in theoretical physics. For decades after Albert Einstein proposed a view of the universe in relativistic terms and nuclear physicists explained the mechanisms of the sub-atomic world by means of quantum mechanics, there have been attempts to merge these two theories. The prize is a theory capable of explaining how very large bodies, and also how very small physical entities, behave. One of the more promising approaches seems to be the *Superstring Theory,* whose aim is to identify the elemental units of structure of the physical world—even smaller than sub-atomic particles such as protons, electrons, and quarks (the building blocks of protons and neutrons).[9]

Why would physicists search for the invisible (as yet unproven and untestable with current technology), very basic components of matter? I fully understand their anguish and the tenacity with which they pursue their aim. Knowledge in the form of concepts, notions, and expressions in language is composed of some basic elements from which all such knowledge emerges and is construed. Like the stem cells in the human body which can develop into different molecules and organs, the strings in the Superstring Theory vibrate in different patterns to form a variety of particles which constitute the sub-atomic structure.[10] Similarly, the fundamental elements of knowledge will produce different types and forms of knowledge—if we can identify them!

If we consider these different approaches and theories, we may find an emerging common thread. Theoretical physicists and neuroscientists have taken up the quest of centuries of philosophers. They pursue the elusive problem of linking those inputs (which we can measure, acknowledge their existence, and are able to experience) with the conceptual constructs that form a reasoned existence in the depths of our mind.

I believe that the model of clustering I describe in the following pages is a form of coexistence of various approaches. To philosophers of knowledge the clustering model offers a compromised solution to the questions: Are there a priori templates of conceptual constructs engraved in our mind? And how do we form these higher-order constructs from senses and experience?

To neuroscientists who are mapping the mechanisms by which the senses are processed in the brain, the clustering model offers an overall framework that explains how the flow and the conjoining of proteins lead to the formation of concepts and higher-order reasoning.[11]

The World According to Initial Clustering

When I first became aware of the Superstring Theory of matter, and the effort by physicists to discover the "theory that explains everything," I was amazed by their vision and the leap in thinking they had undertaken. As I began to compare the elements of their theory to my (yet unformed) view of the structure of knowledge, it occurred to me that as much as the Superstring Theory was revolutionary, it was no more than an extension of what we know, what we had already experienced, and what we have sensed at a given point in our existence.

It dawned on me that there cannot be a phenomenon by which the human mind creates higher-order constructs from *other* higher-order constructs. Rather, the pathway seems to always be from sensorial inputs, enmeshed in accumulated experience—to higher-order constructs. It became ever clearer to me that the mind is unable to replicate higher-order constructs, just as a painter cannot do a painting unless he paints it again, using the basic ingredients such as a canvas, paint, a brush, and his skills. In other worlds, automatic cloning of higher-order constructs, or alternatively, the proto-generation of such constructs from other higher-order constructs, is impossible. I began to realize that there is a given mechanism by which knowledge is created by means of the transformation of lower-order components that cluster and recluster again to form ever more complex and higher-order constructs of knowledge.

This approach, where we descend to the most fundamental unit we can measure, and from it to ascend to the complex phenomenon, is usually described as "reductionism." I prefer to look at it as a more *structuralist* or *mechanistic* approach. But, all these terms describe the method by which knowledge is created by the clustering of sensorial inputs.

The phenomenon or process of clustering occurs in the brain. There are some 100 billion neurons in the two hemispheres of the human brain. Two questions arise: First, what is the significance of this division of the brain to cognition and the formation of knowledge? And second, how does it work? How do the neurons, by interacting and exchanging proteins, cluster the sensorial inputs into higher-order constructs and meaningful concepts?

The Brain and Functionality

Cognitive scientists and brain researchers have explored this first question (duality of the brain's location of specific functions) for some time now.[12] It is my view that the brain is merely a bedrock or a platform in which neurons are able to capture sensorial inputs and to cluster them into the fundamentals of knowledge. To do so would require a platform conducive to this function—hence the division of the brain into hemispheres. Cognition, or the platform for knowledge formation, occurs in the left hemisphere, whereas internal control of bodily functions and physical capabilities occurs in the right hemisphere. Other capabilities are found elsewhere in the brain's geography.[13]

In order for the neurons (whose function it is to process sensorial inputs into clusters and higher-order concepts) to perform their task, they need an adequate platform or *architecture* of physical attributes as well as the biochemical attributes of such a foundational arrangement. This platform combines with their genetic role as processors of sensorial inputs and clusterers of the foundation of knowledge. These requirements for completion of their task require a specialized architecture that cannot be shared with neurons whose task is to control noncognitive functions of the human body and its organs.

Biological systems can be multi-task entities, but cells have a specific task and function. This is analogous to the circumstances surrounding my writing of this book. I sit at my desk, with music playing in the background and a cup of green tea to nourish my thirst. Books, manuscripts, and a powerful personal computer are within reach. All these artifacts make up the architecture of the conditions I need to create this book. But, they are not enough. They are the components of a necessary background within which I can satisfy my inner drive to create and to fulfill my role as a writer. The specialized location of cognition activities in one portion of the brain is not accidental nor insignificant. It is crucial for these cells to function unhindered in an appropriate architecture.

The Major Unsolved Problem in Biology

Francis Crick (1916-2004) co-discovered with James Thompson the molecular structure of nucleic acids (*The Double Helix Structure of DNA*) and the scheme by which replication occurs. Later, Crick (1988, 1995) became interested in neurobiology.[14] Wondering how neurons, by their exchange of biological matter, can give rise to concepts and to consciousness, Crick declared this phenomenon as "the major unsolved problem in biology."

The preoccupation with the way in which *consciousness* is formed in the brain has been a consistent and pesky problem for neuroscientists and philosophers alike. Among their conclusions to date is the belief that although there is a division of the brain into two hemispheres and other selective areas with specific roles in cognition and control, there may be only one consciousness.[15] The formation of representations of nature, such as visual representation of an image, is distributed in various areas of the brain. This provides some measure of efficiency in neuron processing and ensures the storage of what is thus gained in different locations of the brain.

There is currently a quest to understand the working of individual neurons and how they act as groups. Christoph Koch (2004) had estimated that as few as 100 neurons are sufficient to represent an image, such as a facial expression. Thus, if we are endowed with 100 billion neurons, the capacity of the brain to process sensory inputs is extremely high.

How Knowledge is Generated

In the beginning there are sensory inputs and a bunch of neurons which reside in a favorable environment where they can process the sensory inputs. There is an overall agreement among neuroscientists, cognitive psychologists, and philosophers of knowledge that some form of clustering, syntheses, or conjunction will thus take place when conditions are favorable and sensorial inputs are received by the appropriate neurons.[16]

There is the emergence of an initial cluster or a conjoint arrangement of sensorial inputs into a formation that is different from its sensorial components. The question is: how much of a difference and is such a difference instrumental in generating the preliminary form of knowledge? In the previous chapter I suggested that the initial clustering is a clustering of *perceptible distortion* in the set of sensorial inputs and their attributes (context attributes and relation attributes, such as speed, form, distance, direction, etc.). I also suggested that this initial unit of clustered inputs can be named KANE (*K*nowledge b*A*sic u*N*it of *E*xistence).

The attraction of these differentials or perceptible distortions is biochemical. The first attraction is *within* attributes of the sensorial inputs. *Context* attributes (such as speed, time, distance) establish attraction. For instance, a visual input of a substance or figure *moving away* will be attracted to an audible input of the sound also *moving away.* The biochemical formations and exchanges imprinted on and by neurons will thus be similar. The *relation* attributes (such as form, intensity, and continuance) will help to initiate clustering.

The ability to cluster the different perspectives of an event (such as sight and sound) is facilitated by the strength of the senses and their enhanced capabilities to scan the environment with multiple monitoring instruments (the senses), each producing a different yet complementary input about the environment being scanned. Imagine a detective agency or intelligence service employing a variety of sources to collect information from the field or event of interest to them. These sources, albeit different, corroborate each other's findings. Satellites, personal accounts, listening devices, and thermal monitoring are all used simultaneously to gain a description of the event. All these inputs are conjoined in a tapestry of knowledge about the event being scanned.

The Nature of KANEs

Knowledge originates in the brain as a cascade of clustering, starting with the initial clustering of sensorial inputs and their perceptible distortion. This primary knowledge creation has no specific function. It is simply a flow of sensorial inputs that are clustered to create the initial KANEs.[17]

The construct of KANE is the initial perceptible, structural appearance, or emergence of knowledge in the human brain. It probably exists in some physical form in our brain, but it has not yet been identified nor its existence empirically measured. The construct of KANE is an artifact useful in explaining the theoretical passage from sensorial inputs to their clustering into a higher-level of aggregation that can be classified as "knowledge."

Imagine the role of the KANE as that of sub-atomic particles proposed by nuclear physicists as theoretical constructs that help to explain their view of how matter is structured. In the theory of matter where strings are the elemental particles, a string will be a construct similar to KANE: we hypothesize its existence but are as yet unable to measure it empirically.

Without complete understanding of how strings form different structures of matter, we are forced or condemned to make some inferences. In the case of the passage from SIs to KANEs, the process may be somewhat less of a "leap of faith" than with strings (as they vibrate) and electrons or persons they form.

KANEs are formed as the first registry of untested sensorial inputs. This first encounter of sensorial inputs generating what is the first elements of perceptible or registered "knowledge" is the initial transformation from senses to cognition.

So, how is this transformation taking place? There are two possible modes of registering sensorial inputs in the neurons. The first may be an "epigraph"[18] or a chemical inscription (mold or signature) that creates a unique format—similar to the DNA. This unique inscription makes up the KANE. The clustering of KANEs is the mode by which representations are created in the brain and concepts are crystallized into a distinct consciousness. Thus, each clustering of KANEs is a unique *experience,* because of the individualized and unique *epigraphs* by which it is formed. The experience is inscribed in the *memory* section of the brain as a *copy* of the KANE.

The second possibility is the explanation of the "leap" from the sensorial inputs to the formation of concepts and consciousness. By a mechanism still unknown, the brain interprets the sensorial inputs, as a person would need the alphabet to generate complex notions and concepts.[19]

In comparison between the mechanism by which knowledge is formed in the brain and sub-atomic physics, we find that nuclear physicists and even string theorists have not been able to offer a good explanation of the phenomenon of formation of matter. How do the elemental particles of matter (protons, neutrons, electrons) combine to form molecules of distinct properties, such as the elements in nature? The Periodic Table of Elements tells us how many of each particle compose the nuclei of each distinct element—with different physical and chemical properties. What we do not know is how these particles combine to create such elements of matter. Theories abound regarding the means by which such combination occurs, but the precise mechanisms are as yet unknown.

Alas, nuclear physicists indeed resort to the use of external forces, such as gravity, as a force of attraction that keeps the particles together—similar to the *vital force* of biological development. Perhaps the proposition of attraction in the form of electrical charges (inherent in the particles) can also be used to provide some explanation to the attraction of SIs in their clustering effort.

The "Force of Attraction" in the clustering of SIs into KANEs may be the affinity of proteins in the neuron exchange as portals for the signature of SIs. The neural platform allows for the *epigraphs* to be formed by how similar or different the biochemical (amino-acid and protein) signatures of the SIs are.

The Clustering Principle of Knowledge

The path from SIs transmitted (to the platform in the brain where they can be interpreted and clustered) and conceptual thinking and reasoning is made up of *clustering*. Therefore, *knowledge* can be defined as: "the capacity to cluster and thus conceptualize signals from the world external to the focal 'knower'." Knowledge may be defined with the formula:

$$K = C3 \quad \left\{ \quad C = C_1 - C_2 - C_3 \right.$$

where:

K = knowledge

C_1 = clustering

$C_2 =$ **c**omparing

$C_3 =$ **c**onceptualizing

Every living biological organism captures and absorbs signals from its environment. Yet, not every organism is capable of conceptualization, reasoning, and consciousness. The ability to create knowledge depends on:

a. The *richness* of the environment from which signals are received.

b. The *ability to cluster* signals.

c. The *ability to interpret* these clusters by comparing them to existing epigraphs and by conceptualizing them.

There is a belief in the field of information theory and information systems that the quality of information is highly instrumental in developing knowledge. In the framework I am proposing here, the quality of information has a negligible impact on the formation of knowledge, as does the classification into the categories of data 6 information 6 knowledge 6 wisdom.[20] The abilities to cluster signals received from the environment external to the knower and to interpret such clusters are the true determinants of knowledge.

Environments rich with signals allow for a more viable and effective application of the *clustering principle* in knowing organisms. Imagine the extreme case of existing in an environment devoid of any sensorial inputs. There are no sounds, light, changes in temperature, smells, and no other sensations. In this *impoverished* environment, there are no sensorial inputs, hence no clustering and no knowledge that emerges.[21]

Knowledge is therefore defined by the clustering principle. K = C3 defines knowledge, so at each stage of clustering, we have a more complex version of knowledge: how we know and how we make sense of what we know.

How We Conceptualize

KANEs are the first level of clustering of SIs on the road to higher-order concepts. The clustering principle calls for the formation of patterns of cells imbued with epigraphs that are compared to existing clusters previously stored in the operating "warehouse" of memory. Brain researchers have concluded that cognitive activities occur in a specialized area of the brain. Such "centralized reasoning" is a more efficient mode of conducting clustering and conceptualization activities.[22] This is because centralization allows for improved control of the operation—quicker and more accurate clustering than if the brain had to assemble sensorial inputs from their individual locus of activity, and because it allows KANEs to be the first level of reasoning, leading to subsequent clustering in similar forms of the initial KANE.

So, how can we "read" or "interpret" the patterns in the epigraph we call *KANE*? Since each sensorial input is received independently, the initial clustering (or KANE) is the first time the individual SIs are brought together as the preliminary clustering of knowledge. As already stated, there is no previous (a priori) benchmark for comparison, except that which is formed by prior clustering of SIs ("experience").

Consider the concept of *danger.* The pathway from the initial clustering of SIs as KANE is a series of clustering events in which incremental additions are made. There is the clustering of SIs to make the color red, plus a sound (roar), plus a smell—all of which are compared with previous epigraphs in memory: interpreted as *danger.* Such a sentiment is the interpretation of a conceptual construct, from a disparate conjunction of sensorial inputs. Once there is a benchmark of the given arrangement or architecture of neurons (and KANEs) spelling "danger," even the comparison of *one* sensorial input (such as sound or smell or color) is enough to evoke the sensation of "changes." There is an efficiency in our ability to interpret concepts from a very small number of components.[23]

Thus, danger will be construed as a situation that is dangerous *to me* or *another.* This includes, for example, the combination of the concepts of *me, harm to me,* and *outcomes* of harm to me (what danger will do to me if not acted upon). There is a clustering of these clusters, each interpreted per the existing benchmark against which comparisons are made by the brain.[24]

Consider the concept of "justice." The clustering of KANEs that ultimately lead to the emergence of this concept is a path from the inputs of SIs to the formation of a higher-order construct. We do not make a leap of faith from the inputs of SIs to a concept such as *justice.* We are not that smart, nor are we endowed with a gift that allows us to make such a leap.[25]

In my view, the construction of higher-order concepts such as *justice* is a process by which we make incremental additions to an empirical cluster. This is an *extension* or *magnification* of empirical states or events captured by the senses and extended to clustering of the senses. KANEs are clustered in a succession that is originally based on an initial clustering of the primeval experience (the "Big Bang" approach).

So, the concept of justice is the result of clustering of empirical states such as: a human in chains, a human in the mud, a painful personal experience, a similar experience to a relative—all extended and clustered into a comprehensive event that is generalizable as "justice" or its corollary: "injustice."

Concepts as logical constructs are merely extensions and magnifications (through clustering) of sensorial inputs to any future appearance of such events. They may be of two kinds: first, appearance of *same* or *similar* events or clustering of Sis, and second, all other events, similar or not.[26]

KANEs and Cognitive Ingrams

Experience is defined here as a combination of sensorial inputs. These are unique imprints of a code, formed by the clustering of the sensorial inputs. Such imprints are "ingrams," which are ingrained, infused, or written into the cognitive tapestry. This is a combination of "ingrammed" sensorial inputs now stored—as components of knowledge—in the memory in a set of neurons clustered under this unique arrangement. It can be revisited in comparison with new combinations of sensorial inputs.

When a new combination is a match to the existing ingram, the mind "experiences" an event or item of knowledge already "known"—hence, in part, the difficulty in experiencing what others, humans or animals, have known or experienced. In the mind's own cognitive tapestry, there is no viable mechanism to compare the ingrams of others. The closest we can get is a "cognitive affinity" in which we evoke our own ingrams we judge to be similar to those described to us by others, as if they were our own ingrams revisited.

It is therefore impossible to "walk in someone else's shoes." Emotional affinity of sentiments such as compassion and understanding rely not on complete empathy, but on crude projection via one's own comparison with one's own ingrams. The emotional aspect of this phenomenon is a high-level construct of what occurs even with the elemental items of knowledge.

There is also the need to distinguish between the physical aspects of ingrams and the initial generation of knowledge. Cognitive ingrams ("cognigrams") are defined here as the *physical* imprints of the sensorial inputs as they are ready to cluster. These are then stored in memory. Although at present we are not able to isolate and identify these imprints or the biochemistry and the neurons they occupy, I envision that we will be able to do so, just as we have deciphered the genetic code. This is one of the next challenges in biology and brain research.

KANEs are the cognitive (knowledge) aspects or manifestations of these imprints. They are in a form that allows for clustering. KANEs are biologically the same as cognigrams, but they are also clustered. The suggested form in which this works is that a *copy* is made of this ingram (which is stored in memory) when the ingram is to be clustered. This means that cognigrams are not lost when they cluster, and that it is possible to reaccess a cognition for a different cluster. This seems to be a viable explanation for the mind's ability to remember and to reuse sensorial inputs once they have been absorbed.

The above model or scenario also means doubling the space needed to generate this type of creation of cognigrams and memory. Such a need is in line with current theories that "homo sapiens" evolved when their brain capacity increased, thus allowing for knowledge creation and utilization, hence for reasoning. In computer terminology this would mean the *memory* (where cognigrams are stored) and the operating system (where the clustering of KANEs takes place).

The Problem of Continuous Clustering

In this book I advance the model of the structure of knowledge as a series of clustering of sensorial inputs into KANEs, and these cluster into more complex forms, and so on, to the generation of higher-order constructs such as concepts of *justice.* The problem that would be inherent in such a model is whether we can cluster forever, and when clustering will be sufficient so that no additional clustering is needed for the development and fine-tuning of knowledge.

This problem is similar to the *Sorites Paradox* discussed earlier in the book: How many grains of sand form a *heap,* and when would one stop adding grains of sand for a heap to have been achieved? A similar problem was postulated by Karl Marx. He described capitalism as an economic system in which, without proper controls, one individual may be allowed to accumulate all the wealth in the world. In the model of continuous clustering, there are internal controls that prevent clustering forever. These are the "ceilings" embedded in the creation of higher-level concepts.

For example, the rational platform that allows for SIs to be clustered into KANEs, and for these to further cluster into more complex constructs, reaches a saturation point when the concept has been through a certain series of clustering events. Compare this process to the formation of atoms, molecules, and more complex entities in the physical world.

In the case of knowledge, clustering may continue at the level of concepts. For example, consider the clustering of the concept of *deity* and *justice,* forming a more complex concept of a "just God." Perhaps the *range* of clustering in knowledge structuring is much wider than in physical entities, yet the principle of saturation in the seemingly endless process of clustering is valid in knowledge as it is in physics.

In summary, knowledge *is* clustering. Experience and its capturing by sensorial inputs is extended into higher-order concepts via clustering. The model of structure of knowledge described in this book may be called *reductionist, structuralist,* or *mechanistic*—all these terms describe my approach to the initial formation of knowledge in the clustering of sensorial inputs. As rational beings we possess the platform that allows us to cluster SIs into meaningful architectures.

It is the clustering that provides meaning. That is, it is not because we have a built-in a priori notion of relations among sensorial inputs that we are able to create meaningful representations of the clusters—rather, it is through the clustering that such meanings are created.

Science still needs to explain how precisely such transformation occurs at the cellular and molecular levels. The model I present in this book offers a framework in which biochemical processes may be explained.

The Quest for Human Consciousness: Is There a Consciousness, and Does It Matter?

The clustering of sensorial inputs will generate knowledge. Previously I also discussed the issue of qualia, or the experiences as they are felt by the knower. In addition to the generation of knowledge, there may also be a phenomenon beyond the biochemical exchanges in the brain. Such a phenomenon would be a form of "thought" or a *mental* construct, beyond the physical aspects of what constitutes the human brain. The quest for such a metaphysical, or mental, construct has been explored for many years, perhaps centuries. It is known as the mind/body problem (Humphrey, 2000), or the quest for the human conscience.[27]

There is a lack of agreement among scholars on the definition of consciousness and its structure or form. The construct of consciousness is generally considered a complex or "mongrel" concept. This leads to what some scholars call "the problem of consciousness" or the "problem of mind-body" phenomenon (e.g., Antony, 1999; Marcel & Bisiach, 1992).

The issue or problem of the existence of consciousness has been attracting the attention of scholars and philosophers for at least a century, with a renewed effort based on modern scientific methodologies and advances in related disciplines such as psychology and brain sciences.[28]

Consciousness has also been classified into four major types. "Phenomenal consciousness" is usually defined as the experience of a human being in terms of the senses, feelings, and thoughts (see Chalmers, 1997).

Another type is "access consciousness," which can be defined as the use of perception or perceptual ability in guiding the individual's cognitive processes and the individual's action. This definition extends the notion of consciousness from the personal experience to the levels of thought and action. Another form of consciousness is the ability to have complex or compound thoughts, leading to the higher-order form of "self-consciousness." In this type of concept, the human being is able to engender the notion of self as an independent entity. It is akin to "leaving one's body" and perceiving of the self from an objective perspective—outside the physical domain of the self (e.g., Dennett, 1991; Blackmore, 2003; Donald, 1991).

Such 'self-awareness' is an advanced ability which humans seem to possess, and which eludes other species. Much has been heralded about this peculiar human cognitive ability. It has been postulated to be the basis for human reasoning, human ethics, and human existence. The philosopher Immanuel Kant had argued that "consciousness" is an epistemological necessity, and we must accept its existence regardless of any test we could apply to verify its existence. Rene Descartes said as much in his famous "cogito ergo sum" (I think, therefore I am). This means that in order to be able to conduct reasoned thoughts and interface, humans must first of all recognize themselves as thinking beings—that is, have self-awareness and recognize their own consciousness.

But, here again we are faced with philosophical maneuvering which can be simply summarized in two distinct frameworks to what constitutes consciousness. One is the contention that "consciousness" exists in a transcendental form outside the physical compounds of human existence. The second is the argument that consciousness is a form of higher-level construct of human thinking, namely, a construct with the cognitive content of morality, existence, high-level experiences, and reasoning—including what may well constitute the essence of our humanity.

However we position the notion of consciousness on the continuum between experience and the extra-physical existence as a separate mental entity, the dispute over what exactly is human consciousness will remain unresolved. Within the framework of this book, considering consciousness as a form of knowledge seems a more attractive attempt to focus the concept, thus avoiding the need to venture into several types and forms of this elusive concept.

If, as I propose below, consciousness is a form of knowledge, there is no need to assert that it is an "epistemological necessity," nor that it is perhaps a cultural construct or a very personal form of experience. There is no further need to view consciousness as an obligatory phenomenon whose existence is essential to being human and to having the ability to reason and to create and utilize knowledge. This means that one must not necessarily have the ability to experience oneself in order to create and generate knowledge. Rather, the direction may be the other way: one will experience the notion of self as a result of the ability to generate and utilize knowledge.[29]

Aside from the ethical, religious, and cultural connotations and consequences of the affirmation of consciousness as a phenomenon of value, its epistemological necessity only matters as far as it is an item of knowledge engendered by the mind, with the same elemental components of any other item of knowledge which coalesces and is clustered to form higher-order constructs.

Consciousness as Knowledge

The notion of consciousness is a good example of the position of this book in the interdisciplinary search and debate. This book mediates between the effort of brain and cognitive scientists (who explore the elemental, cellular, and molecular biology and chemistry of the brain), and the philosophers and cognitive scholars who explore the higher-order constructs and notions of cognition and reason.

Consciousness is a form of knowledge derived from the clustering of sensorial inputs into KANEs, which happen to engender the higher-order construct of "knowing oneself." The process of generation of knowledge is very similar to that of engendering knowledge about space, distance, colors, and other attributes of the environment external to the knower. In this very particular case, the architecture of clustering happened to focus on attributes of the self.

In a way this model is again an interpretation of Rene Descartes' (1596-1650) famous conclusion: "Cogito ergo sum" (I think, therefore I am). Although this may not have been Descartes' meaning, his declaration may be viewed as: "I generate knowledge about myself, therefore I know I exist, therefore I am." Within the constraints of this book's journey (in search of the elemental building blocks of knowledge), the ability to create knowledge about one's self is just another mode of creating knowledge. Sometime during the evolutionary process, humans had arrived (by accident or mistake) through a spike in the evolutionary journey at an architecture of KANEs which allowed them to generate a higher-order notion of their presence within their environment.[30]

Comparing human ability to identify their own existence with other species is at best a view anchored solely in ethics and religious beliefs. Human beings exist in other forms of skills and knowledge, such as certain uses of tools, and reasoning beyond immediate experiences—that is, generating knowledge in the form of higher-order constructs which are separated from the immediate sensorial inputs in time and topic. Although I recognize the ontological and ethical implications that may result from the notion of consciousness, it remains just a form of knowledge created by the human mind.[31]

References

Antony, M. (1999). Outline of a general methodology for consciousness research. *Anthropology and Philosophy, 3*(2), 43-56.

Bergson, H. (1998). *Creative evolution.* Mineola, NY: Dover.

Blackmore, S. (2003). *Consciousness: An introduction.* New York: Oxford University Press.

Buzaglo, M. (2002). *The logic of concept expansion.* New York: Cambridge University Press.

Capra, F. (1997). *The web of life: A new scientific understanding of living systems.* New York: Anchor Books.

Capra, F. (2002). *The hidden connections: Integrating the biological, cognitive, and social dimensions of life into a science of sustainability.* New York: Doubleday.

Chalmers, D. (1997). *The conscious mind: In search of a fundamental theory.* New York: Oxford University Press.

Chen, C., & Kulijis, J. (2003). The rising landscape: A visual exploration of superstring revolutions in physics. *Journal of the American Society for Information Science and Technology, 54*(5), 435-446.

Churchill, F. (1969). From machine-theory to Entelechy: Two studies in developmental teleology. *Journal of the History of Biology, 2*(2), 165-185.

Churchland, P. (1988). *Neurophilosophy: Toward a unified science of the mind/brain.* Cambridge, MA: MIT Press.

Clarke, E., & O'Malley, C. (1968). *The human brain and spinal cord: A historical study illustrated by writings from antiquity to the twentieth century.* Berkeley: University of California Press.

Close, F. (2000). *Lucifer's legacy: The meaning of asymmetry.* New York: Oxford University Press.

Collins, H. (1992). Will machines ever think? *New Scientist, 1826*(June 20), 36-40.

Cooper, D., Mohanty, J., & Sosa, E. (Eds.). (1999). *Epistemology: The classic readings.* London: Blackwell.

Crick, F. (1988). *What mad pursuit: A personal view of science.* New York: Basic Books.

Crick, F. (1995). *Astonishing hypothesis: The scientific search for the soul.* New York: Scribner.

Damasio, A., & Grosset, D. (1995). *Descartes' error: Emotion, reason, and the human brain.* New York: HarperCollins.

DeLessemme, A., & DeDuve, C. (2001). *Our cosmic origins: From the big bang to the emergence of life and intelligence.* New York: Cambridge University Press.

Dennett, D. (1991). *Consciousness explained.* New York: Little, Brown, and Company.

Donald, M. (1991). *Origins of the modern mind: Three stages in the evolution of culture and cognition.* Cambridge, MA: Harvard University Press.

Edelman, G. (2004). *Wider than the sky: The phenomenal gift of consciousness.* New Haven, CT: Yale University Press.

Edelman, G., & Tonomi, G. (2001). *A universe of consciousness: How matter becomes imagination.* New York: Basic Books.

Greene, B. (2004). *The fabric of the cosmos: Space, time, and the texture of reality.* New York: Alfred A. Knopf.

Humphrey, N. (1999). *A history of the mind.* New York: Copernicus Books.

Humphrey, N. (2000). How to solve the mind-body problem. *Journal of Consciousness Studies, 7*(4), 5-20.

James, W. (1904). Does consciousness exist? *Journal of Philosophy, Psychology, and Scientific Method, 1*(1), 477-491.

Jaynes, J. (2000). *The origins of consciousness in the breakdown of the bicameral mind.* Boston: Mariner Books.

Kauffman, S. (1993). *The origins of order: Self organization and selection in evolution.* New York: Oxford University Press.

Kearney, A., & Kaplan, S. (1997). Toward a methodology for the measurement of knowledge structures of ordinary people: The conceptual content cognitive map (3CM). *Environment and Behavior, 28*(5), 579-617.

Koch, C. (2004). *The quest for consciousness: A neurobiological approach.* Englewood, CO: Roberts and Co.

Ledoux, J. (2003). *Synaptic self: How our brains become who we are.* New York: Penguin Books.

Little, D., & Thulborn, K. (2005). Correlations of cortical activation and behavior during the application of newly learned categories. *Brain Research: Cognitive Brain Research, 25*(1), 33-47.

Lynch, G., MacGaugh, J., & Weinberger, N. (Eds.). (1984). *Neurobiology & learning and memory.* New York: Guilford.

Marcel, A., & Bisiach, E. (Eds.). (1992). *Consciousness in contemporary science.* Oxford, UK: Clarendon Press.

Marcus, G. (2003a). *The algebraic mind: Integrating connectionism and cognitive science.* Boston: MIT Press.

Marcus, G. (2003b). *The birth of the mind: How a tiny number of genes creates the complexities of human thought.* New York: Basic Books.

O'Reagan, K., & Noe, A. (2001). A sensorimotor account of vision and visual consciousness. *Behavioral and Brain Sciences, 24*(5), 979-980.

Ornstein, R. (1992). *Evolution of consciousness: The origins of the way we think.* New York: Simon & Schuster.

Ornstein, R. (2000). *New world new mind: Moving toward conscious evolution.* Cambridge, MA: Malor Books.

Place, U. (1956). Is consciousness a brain process? *British Journal of Psychology, 47*(1), 44-50.

Posner, M. (1973). *Cognition: An introduction.* Glenview, IL: Scott Foresman.

Quirk, T. (1990). *Bergson and American culture.* Chapel Hill: University of North Carolina Press.

Ridley, M. (2001). *The cooperative gene: How Mendel's demon explains the evolution of complex beings.* New York: The Free Press.

Rosch, E., & Lloyd, B. (Eds.). (1977). *Cognition and characterization.* Hillsdale, NJ: Lawrence Erlbaum.

Rose, S. (Ed.). (1999). *From brains to consciousness? Essays on the new sciences of the mind.* Princeton, NJ: Princeton University Press.

Shafto, M. (Ed.). (1986). *How we know.* San Francisco: Harper & Row.

Simon, H. (1996). *The sciences of the artificial* (3rd ed.). Cambridge, MA: MIT Press.

Velmans, M. (2000). *Understanding consciousness.* London: Routledge.

Velmans, M. (Ed.). (1996). *The science of consciousness: Psychological, neuropsychological, and clinical reviews.* London: Routledge.

Webster, L., & Goodwin, B. (1996). *Form and transformation: Generative and relational principles in biology.* New York: Cambridge University Press.

Zimmer, C. (2004). *Soul made flesh: The discovery of the brain and how it changed the world.* New York: The Free Press.

Endnotes

[1]   There are specific and esoteric terms used by scholars in various disciplines related to mind, knowledge, and cognition. By "structuralist" approach I refer to the reliance my model makes on the conjoining architecture and clustering effort by which the mind arrives at a meaningful structure of sensorial inputs. This approach may be contrasted with approaches that advocate a priori existence of frameworks of reason and logic—upon which sensorial inputs may be imprinted and knowledge on concepts thus are gained. See, for example, Buzaglo (2002), Humphrey (1999), and Cooper, Mohanty, and Sosa, (1999).

[2]   Also see Churchland, P. (1980). A perspective on mind-brain research. *Journal of Philosophy, 77*(4), 185-207. For more recent research, see Little and Thulborn (2005) and Rose (1999).

[3]   For further readings about consciousness and related philosophical explorations, see Dennett (1995). The discussion of consciousness is given in this chapter.

[4]   I cite a select group of scholars, chosen by my view of the relation of their work to my framework, and my perceived importance of their work. Readers may wish to consult a host of other scholars. In this topic, readers may consult Shafto (1986) and other cognitive psychologists.

[5]   Since Bergson there have been several schools of thought in philosophy of knowledge and a move toward explorations of neuropsychology and neurobiology. The problem remains unsolved. A colleague suggested that this problem is the road "from the Amoeba to Einstein," as proposed by evolutionary epistemologists (Chapter VIII). The problem, as I state it, is more focused: the road to reason is the path from sensorial inputs to the formation of concepts and the elementary unit of what constitutes knowledge. Instead of the progression advocated by evolutionary epistemologists, I am advancing the notion of "clustering."

[6]   In 2004, the Nobel Prize for Medicine (Physiology) was awarded to Richard Axel and Linda Brek for their work on the pathways from the nose to the brain. Their work helped to explain the olphatic sense and how such sensorial input is processed in the brain. This research is illustrative of the effort to map the biochemical processing of the senses in the brain, and the proteins that conduct such processing.

[7]   In Chapter IV I discussed the issues of *qualia* that O'Reagan and Noe (2001) had tackled.

[8]   O'Reagan and Noe (2001) defined the general mode of visual consciousness as "just the higher-order capacity to exercise such mastery" over the sensorial inputs. This view corresponds to the clustering mode that I present in my model.

[9]   See the discussion in Chapters II and III. For a popular description of this theory, see Greene (2004). The reader who wishes to explore the development of the Superstring Theory should consult Chen and Kulijis (2003). In 2004 the Nobel Prize for Physics was awarded to David Gross, David Politzer, and Frank Wilczek for their work on the "strong force" within the atom that binds quarks. The Swedish Academy, in justifying the prize, said: "Their work helped science get closer to a theory for everything." This is the term often used to describe the Superstring Theory. Also see Chapter III.

[10]   The Superstring Theory suggests that matter is originally formed when strings or filaments of energy vibrate to create different types of matter with their own attributes, thus forming quarks or electrons. This is an approach to the sought-after unified theory of the universe.

[11]   Even the Superstring Theory is anchored in human experiences with the senses. The theory is sometimes described as a leap into ideas of spatial arrangements and multiple dimensions beyond the current human ability to test and measure them. Yet, one key property of the strings is their ability to vibrate, like the string of a musical instrument, whereby the harmonics of their vibration produces characteristics of matter. Vibration and harmonics are phenomena captured by human senses. However visionary this theory may be, it can only be described and explained with the way sensorial inputs are processed and transformed into higher-order entities—the basic elements of matter or the elements of knowledge.

[12]   There is a vast and growing literature. For a short list of the more salient contributions, see Koch (2004), Marcus (2003), Edelman (2004), and Ledoux (2003).

[13]   The discourse on brain physiology and its impact of knowledge formation is given in a very brief and summarized format, not as a scientific treatise. Among others, see Edelman and Tonomi (2001), Humphrey (2000), Lynch et al. (1984), and Churchland (1988).

14   In 1962, Crick and Watson shared the Nobel Prize with Maurice Wilkins (1916-2004) for "the three-dimensional molecular structure of the DN...the substance carrying the heredity" (from the Nobel Prize in Physiology or Medicine presentation).

15   In this book I explore the emergence of *knowledge* from the clustering of sensorial inputs, not the formation of *consciousness*. Although these may be different phenomena, the process of their generation may be quite similar.

16   The problem of mind-matter and the emergence of consciousness from the mere biochemical exchanges in the brain has been extensively studied, although with little hard evidence as to how this works, and with a large dose of speculation. See, for example, Damasio (1995), Blackmore (2003), Marcus (2003), Edelman (2004), and Edelman and Tonomi (2001).

17   As I am sitting at my desk, looking out on the magnificent old oak tree adjacent to the house, in the process of shedding its leaves for the forthcoming winter, it seems to me that I might be a cog in the wondrous machination of scientific serendipity. As the story goes, I look forward to such a revealing occasion, as experienced by my namesake Eli Whitney, to the cat and the hens and a fence, and the idea of a cotton gin. It may be that the phenomenon of sensorial inputs to cognition is as simple as ABC, or as complex to defy my limited abilities of clustering the known into the mysterious working of the human mind.

18   An epigraph is an engraved inscription. This term would best describe the engraving upon the neurons of the individual and unique inscription of the perceptible SIs so as to form a KANE.

19   Earlier in the century some biologists had argued that a special force may exist in nature, so that living organisms are driven by a vital principle of life that facilitates development of living beings. Hans Driesch (1867-1949), for example, named this force "entelechy," after the Aristotelian concept of self-accomplishment and growth—from the acorn to the mighty oak. Thus, the mechanistic approach to growth and development is enhanced, or even replaced, by the *vital* force—external to the process of growth—that may offer an explanation to why and when growth occurs from cells to higher-order organisms. A similar perspective may be applied to the task of explaining how SIs "develop" into higher-order concepts and consciousness. Just as *vitalism* is not an acceptable explanation for the development of living organisms, so

is an unknown and external factor not acceptable as an explanation to how SIs "grow" into conceptual thinking. The mechanistic perspective must prevail. (For additional readings, see: Capra (1997, 2002), Quirk (1990), Churchill (1969), Webster and Goodwin (1996), and Collins (1992).

[20]   In the Jewish religious and mystical tradition, there is a Hassidic movement called "HABAD." The name is an acronym of the Hebrew terms: "*H*ochma" = wisdom; *B*ina = knowledge with understanding/intelligence; *D*aat = knowledge. The elements of the acronym may be viewed in a descending order, starting with simple knowledge and advancing to wisdom. Such a trajectory is very similar to the clustering of elemental knowledge into higher-order concepts which may be considered the components of intelligence and wisdom. This, however, is different from the path data 6 information 6 knowledge, as the starting point in the pathway is *knowledge.*

[21]   In Chapter XII will consider this thesis by advancing a definition of life that equates it with the ability to cluster sensorial inputs into knowledge. As a fetus is able to absorb sensorial inputs *and* to cluster them into knowledge, it is a living *and* knowledgeable (armed with awareness) organism.

[22]   I am introducing organizational and managerial principles and terminology of efficiency of operations to explain the working of the "reasoning" function of the brain.

[23]   See, for example, Kearney and Kaplan (1997). They proposed cognitive maps as mental models that represent the structure of knowledge. They describe internal representations, hierarchical structures, and sequential coding as modes of forming cognitive maps. They define these maps as: "hypothesized knowledge structures embodying people's assumptions, beliefs, "facts," and misconceptions about the world" (p. 580).

[24]   The reader may wish to consult Posner (1973) and Rosch and Lloyd (1977).

[25]   The reader may wish to consult different perspectives and approaches, such as Edelman (2004), Zimmer (2004), Koch (2004), Ledoux (2003), and Blackmore (2003).

[26]   This knowledge of concepts is diffused among humans by the rise of language. We are able to thus communicate at the conceptual level because we develop these concepts by means of equal if not very similar

platforms in our brains. Hence, the role of language may be reducible to the representation of commonalities in clustering individual experiences. See, for example, Ridley (2001), DeLessemme and DeDuve (2001), and Close (2000).

[27]   There is extensive literature on the subject, primarily in philosophy, metaphysics, and related disciplines. For an initial exposure to this literature, see Jaynes (2000). Ornstein (1992, 2000). Blackmore (2003). and Velmans (1996, 2000).

[28]   For some early investigations, see James (1904).

[29]   Other species, such as birds and dolphins, have a much more advanced knowledge of direction and position in their environments than humans.

[30]   For additional readings, see Kauffman (1993), Place (1956), and Simon (1996).

[31]   At some point in the future, we will be able to track and identity the formation of clusters or KANEs which will denote the notion of consciousness. Similarly, by having such capability we will be able to tell what a person is thinking and, in other uses, whether the person is telling the truth. Other uses would be an enormous improvement on our measures of intelligence (IQ) and emotional intelligence (EQ). This will truly be a "brave new world."

Chapter VII

How We Organize What We Know

The Unity in Modeling

In developing the model of the structure of knowledge, I embraced the risk of wandering off into the wilderness of marginal, perhaps inconsequential, modeling. This first part of the book addresses the structure of knowledge, whereas the second part deals with how knowledge grows, progresses, and advances. I had written several papers about how I believed human knowledge progresses. I had flatly rejected the evolutionary model of knowledge growth and progression. My views had congealed in the form of a different, clear, and consistent model.

The structure of knowledge was a different story. I had begun with some rough ideas of how knowledge begins and how it is structured. A constant companion has been my fear that I would end up with a *duality* rather than *unity* in the modeling of structure and progress of human knowledge.

This intellectual challenge resembled the digging of a tunnel into a mountain. One best approaches such an undertaking by starting to dig from both sides of the tunnel—in the hope that one's calculations are accurate and the two teams will meet halfway exactly as planned. The emergent "clustering principle" in the structure of knowledge had to be in line with my model of progress which stresses cumulation. As I explain further in Part II, I am delighted to inform the reader that the two models meet exactly as hoped.[1]

Structure and Balance

The key for the principle of clustering as the guiding principle of knowledge generation and its structure has been simplicity. This is a way of thinking about a complex phenomenon by taking the simplest approach. Albert Einstein presumably once declared: "Make everything as simple as possible, but no simpler."[2] Einstein himself approached the problem of the aether (through which light travels) in the simple way of declaring that there was no such entity as the aether and that speed of light is a constant in the universe, regardless of the medium through which it travels or the position and motion of the observer.

One would assume that as knowledge is formed by continuous clustering, there would be some points along this process in which a *balance* would be attained. Such possible homeostasis would mean a period in which the knowledge generated up to that point can be functionally utilized, measured, and accessed without further additions or continuing clustering. This would be a situation similar to Gould's notion of "punctuated equilibrium" in biological evolution: periods of evolutionary activity followed by periods of relative stability and equilibrium.

In my effort to exert unity in modeling between structure and progress of knowledge, I opted for the rejection of attempts to uncover, identify, and understand the existence of balance or situations of homeostasis in the generation and structure of knowledge. Rather, I started with the notion that there is no perceptible nor measurable balance in the continuous clustering of knowledge. "Continuous clustering" is enmeshed with "continuous cumulation," which is the principle by which knowledge progresses. Within the unity in modeling, there is no room for an inherent phenomenon of balance.

The seemingly active situation of balance or equilibrium in how knowledge emerges and grows is merely the bias of the observer. Clustering and cumulation are ongoing regardless of the observer's position or perspective. Hence, this is similar to Einstein's notion of the speed of light as a constant in the universe. The emergence and growth of knowledge by clustering and cumulation are a constant—in the sense that they are independent of the observer. This conclusion is based, of course, on the assumption that there is more than one "creator" of knowledge. Each generates knowledge from sensorial inputs, so that through clustering by *each,* and cumulation by *all,* knowledge grows and progresses.

Early in the preparation of this book, it had occurred to me that in order to better understand human knowledge, we need to start at the elemental level and work from the bottom up. It is in human nature to utilize the existing, measurable framework and to work from these in the search for answers. I decided to challenge conventions and to address knowledge not as the phenomenon we recognize in ourselves and in our organizations, but rather as an unknown phenomenon, open to a fresh and unriddled outlook on its origins and progress.

By doing so, and by following this unorthodox and bottoms-up reductionist approach, I was able to construct a comprehensive framework of human knowledge. This effort goes well beyond the current state of the taxonomic state of knowledge research. In my understanding, it is the first attempt to frame a broad model of the generation, structure, *and* progress of knowledge.[3] I believed that if we had simply remained at the macro level of how knowledge appears and is utilized, we would be going in circles on which categories to use and how higher-level concepts inter-relate and interact.

The Four Paradoxes

In a broader sense, the way we understand and organize knowledge in our life is anchored in four inter-related paradoxes. The term "paradox" means a mode of thinking or a statement about something that runs contrary to what people expect and to common sense, yet it might well be true. These paradoxes are: (1) the Sorites Paradox, (2) the Paradox of Knowability, (3) the Paradox of Zero Knowledge, and (4) the paradox of complete or total knowledge.

These paradoxes provide us with the following benefits. First, they *clarify concepts* and make them more transparent. Secondly, they *set boundaries* to the construction of knowledge and the conceptual frameworks that it forms. The paradoxes of zero and complete knowledge are a good example, as they set the boundaries of where knowledge may reside. Thirdly, the paradoxes *help to explain limitations and strengths* of the approach to knowledge I have advanced in this book. Finally, they *offer direction* for further exploration of the nature of knowledge.[4]

The *Sorites Paradox* concerns the point at which a "heap" is considered a heap, or how many grains of a substance make a clustered entity. In the case of knowledge, the clustering of KANEs generates more complex entities—each still constituting knowledge. Unlike the distinction between, for example, grains of salt and a heap made of assembled grains, the clustered entity increases in complexity but does not change its nature.

Fitch's *paradox of knowability* claims that if all truths are knowable, then it also means that all truths are known. This is so because, as Frederic Fitch suggested in 1963, if we accept the proposition or the principle that all truths are knowable, then we must also accept the proposition that we cannot know an unknown truth. In other words, since humans are human, not all-knowing (omniscient), then it becomes impossible for us, as humans, to know *all* that is true.

Notwithstanding the logical analysis that leads to Fitch's paradox, the extension of the knowledge we do possess to *all* knowledge and all that is true remains a philosophical argument.[5] In my model of the structure of knowledge, the clustering of sensorial inputs and subsequently KANEs does not imply that such knowledge we generate is a true representation of reality. Within the limits of clustering, there will be knowledge that has not been clustered or captured by our sensorial inputs. But, the model of the structure of knowledge offered here considers the knowledge that *is* generated from sensorial inputs, not all available knowledge in the universe of *all* knowers. I am specifically concerned with the knowledge we generate from our sensorial inputs and reject its extension to all possible knowledge which could be generated but has not been so created.

The paradox of *zero knowledge* argues that if there is zero or no knowledge, then it must be known that there is zero or no knowledge. Then, if it is so known, there is *some* knowledge, not zero. In other words, we can never *know* that there is zero knowledge. Like matter, there will always be some knowledge. The point at which there is zero knowledge is also the beginning of the knower—as knower. Hence, knowledge can only exist in relation to and in the presence of the knower.

I ask: how can there be knowledge without the knower and a set of criteria that establish the absence of knowledge? The discussion along these lines will lead to a process of *reductio-ad-infinitum,* to the *original* mechanisms that recognize knowledge and has the *original* set of criteria.

The paradox of zero knowledge provides a lower limit for the appearance of the knower—as knower. The knower may have "no knowledge" of a particular topic or problem, or phenomenon—but the knower is never without *any* knowledge. Even to recognize the absence of knowledge in a particular area is to possess knowledge that allows the knower to establish and to recognize ignorance.

This discussion is not an argument in support of consciousness. In order for such an argument to be accepted within the framework of this book, *initial* knowledge would always have to be that of consciousness. This is not the case. Initial knowledge may be the clustering of sensorial inputs to form knowledge about the external environment without the notion of knowing oneself. A newborn may "know" the mother as a source of food without the conception of the self.

Finally, the paradox of *complete knowledge* claims that if all knowledge is known, then the knower (who is at this juncture omniscient) has infinite knowledge, so that knowing that one has complete knowledge goes beyond completeness. In other words, complete, infinite, or total knowledge is unknowable.

The notion of complete knowledge is related to the concept of "clustering ad-infinitum." As we cluster KANEs into ever more complex architectures of knowledge, we are eventually striving toward achieving complete knowledge. The question is: how more complex can we get? As we construct more conceptual knowledge architectures, we are approaching the conceptualization of complete knowledge, also translated as *foreknowledge* and *after knowledge.* That is, knowledge beyond the confines of time, attributable only to the omniscient. This leads to the hypothesis that if such complete knowledge can be achieved, it is nested in a deity. Only God possesses all possible knowledge of what was, what is, and what will be.

However, if we define knowledge as the clustering of our human sensorial inputs, then complete knowledge is the totality of all *human* knowledge. By extension, God would have his own knowledge based on the clustering of his own sensorial inputs. Philosophical and theological discussions would argue here that as a spirit, God *is* knowledge, hence its knowledge is purely conceptual—devoid of sensorial inputs. To avoid such discussions, let me state that the paradox of complete knowledge is a good indicator of the logical existence of God in the universe. Once a distinction is made between human knowledge (based on sensorial inputs) and the complete knowledge which only a deity may possess—there is no doubt that such an entity is a logical necessity.[6]

Between Zero and Completeness

Human knowledge resides between zero and complete knowledge. These extremes are the boundaries where knowledge would accumulate to form a distinctive entity. However, in the model of structure I present here, human knowledge is a function of the knower, and a result of the clustering of individual sensorial inputs.

So, even though we are able to communicate our sensations and the knowledge we generate, human knowledge is our individual experience and a personal phenomenon. The structure of knowledge presented here refers to the elemental units of knowledge that are within the person. The totality of knowledge created by people is not an independent entity. Thus, the "state of knowledge" in any given area of human endeavor is the sum total of what humans know in that area and are able to communicate and to exchange with each other. There is not an ontological distinction of a body of knowledge.

Why? If we assume the existence of such a body of knowledge, we should also have the origin of such an entity outside the existence of the knowers. When does such an entity start and at which point beyond "zero knowledge" does it become a "body" of knowledge?[7]

Managing and Measuring Knowledge

We have already ascertained that the structure of human knowledge increases in complexity, but it does not become too complex for people to manage it or to communicate it to other people. We form knowledge architectures at the level of inter-related concepts which are woven into these manageable structures. Concepts do not beget concepts. They are the product of clustering.

Figure 1. Measures (attributes) of knowledge

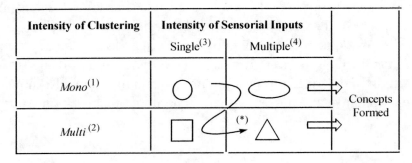

Intensity of Clustering	Intensity of Sensorial Inputs	
	Single[3]	Multiple[4]
Mono[1]		
Multi [2]		

How do we measure knowledge? There are two key measures (or "attributes") of knowledge: (1) the intensity of clustering, and (2) the intensity of sensorial inputs. This is shown in Figure 1.

As we cluster SIs and form even more complex concepts, we are also clustering these diverse clusters of complex knowledge. In the space between zero knowledge (infinitely small) to complete knowledge (infinitely large), we generate volumes of knowledge that are measurable and manageable, so they can be stored, retrieved, and communicated.

A Periodic Table of Knowledge

In the architecture of elements of matter, it is possible to form a periodic table in which elements can be shown and predicted. Is such a table possible for knowledge? Based on the clustering of SIs and KANEs, is it theoretically possible to predict the elements of knowledge that could or would be created by different clustering architectures?

I believe that this is possible, without resorting to considering knowledge an independent entity outside the knower. Different architectures of clustering in a given set of SIs and KANEs may generate different types of knowledge, and a different set of concepts can thus be created. This is a problem that should occupy knowledge sciences for years to come.

Rationality Revisited

How do we organize what we know? We do so based on the rationality of the knowledge we generate and its rational propensity to be condensed into databases and knowledge warehouses.[8] But, what is the relationship between knowledge and rationality?

The question we should ask is: how does what we know lead to rationality? We are bounded or limited by the constraints we have in processing knowledge (bounded rationality) and by the notion I advanced in this book that knowledge *is* clustering of SIs and KANEs, and that concepts do not beget concepts.

Will the clustering of elements of knowledge lead to rationality? Is there such a construct as "knowledge-based rationality (KBR)"? And does rationality increase or improve with additional knowledge? These questions are based on the link between the generation of knowledge and its utilization in the formation of rational constructs, propositions, and architectures of what we know. If this is so, then the link between clustering and rationality will depend on the quantity of knowledge, on its richness, and on its quality.[9]

In the model of the structure of knowledge proposed in this book, I distinguish between the creation and structuring of knowledge—and its utilization in cognition, rationality, and problem solving. The model concentrates on the initial generation of knowledge, prior to its transformation into science, social knowledge, and other advanced forms of formatting what we deem to be "knowledge." This distinction between the structure and utility of knowledge will also be of paramount importance in Part II of this book—as an explanatory variable of how knowledge progresses.

Here, suffice it to state that the creation of knowledge by means of clustering of SIs and KANEs does not add to increased rationality in the manner in which we utilize such knowledge. The accumulation of knowledge is done by continually clustering "from scratch."[10] We generate knowledge by continuous clustering, thus accumulating knowledge by comparison, but not as raw material upon which a "body" of knowledge is built. The process of knowledge generation starts anew on a continuous basis, as a renewable continuous clustering.[11]

This perspective of continuous generation of knowledge gains support from recent findings of the functioning of the human brain. There is increasing evidence that the brain is highly efficient in the location of its functions and in processing of inputs. The brain operates on the basis of comparison by proximity, hence the need to employ distinct and well-focused areas of activity.[12]

Continuous Clustering and Networking

The model of continuous clustering and the proposition that the brain operates in a very efficient spatial manner brings into question the use of neural networks and parallel processing as emulators of the brain's mode of knowledge creation and processing. The rapid growth in this field of neural networking has, by and large, masked its weaknesses.[13] Neural networks are an illusion where the structure of knowledge is concerned.

Neural networks do not allow for the limits of complexity that characterize the structure of knowledge. In generating knowledge, one creates a tapestry, woven by criteria for clustering—not by networking with unlimited expansion.[14] Networking is based on the premises that: (1) the elements in the network join at different times and for different reasons or criteria, (2) the guiding principle in the network is *not* clustering, and (3) there is an exchange among the elements or members of the network anchored in a structure of parallelism or activities that occur in parallel.[15]

Although neural networks, like in continuous clustering, "start over" at each juncture, they fail to imitate human neurons. They are based on some advances to digital and analog computing, where nodes are activated with different weights or electrical impulses. This is a far cry from the ability of neurons to absorb inputs and process them by clustering and the creation of higher-order constructs of knowledge.

Hence, the brain does not need to network nor to parallel-process inputs. Knowledge is created by clustering and comparison—in a mode that is both localized and efficient.[16] Networks have an important role in diffusion and transfer of knowledge—not in its creation.[17]

In summary, we generate knowledge by continuous clustering of SIs and KANEs. How do we gather, accumulate, and grow such knowledge? What is the model of growth of knowledge? Can we predict what knowledge will or should be generated—as physical scientists do with the periodic table of the elements? With these mighty questions we embark on the journey to the second part of this book: the progress of knowledge.

References

Agamben, G., & Attell, K. (2004). *The open: Man and animal.* Stanford, CA: Stanford University Press.

Boutsinas, B. (2004). Incorporating common sense into a machine learning system. *International Journal of Computational Cognition, 3*(3), 1-18.

Buchanan, M. (2003). *Nexus: Small worlds and the groundbreaking theory of networks.* New York: W.W. Norton & Company.

Capra, F. (1997). *The web of life: A new scientific understanding of living systems.* New York: Anchor Books.

Close, F. (2000). *Lucifer's legacy: The meaning of asymmetry.* New York: Oxford University Press.

Collins, H. (1990). *Artificial experts: Social knowledge and intelligent machines.* Cambridge, MA: MIT Press.

Davidson, D. (2004). *Problems of rationality.* New York: Oxford University Press.

Domhoff, W. (2002). *The scientific study of dreams: Neural networks, cognitive development, and content analysis.* Washington, DC: American Psychological Association.

Elster, J. (1989). *Solomonic judgments: Studies in the limitations of rationality.* New York: Cambridge University Press.

Galison, P. (1999). *Images and logic: A material culture of micro-physics.* Chicago: University of Chicago Press.

Garcke, H., Preuser, T., & Rumpf, M. et al. (2001). A phase field model for continuous clustering on vector fields. *IEEE Transactions on Computer Graphics and Visualization, 7*(3), 230-241.

Gigerenzer, G., & Selten, R. (Eds.). (2002). *Bounded rationality: The adaptive toolbox.* Cambridge, MA: MIT Press.

Girolami, M. (1999). *Self-organizing neural networks: Independent component analysis and blind source separation.* New York: Springer-Verlag.

Glanvill, J., & Fabian, B. (Eds.). (1979). *Plus ultra or the progress and advancement of knowledge since the days of Aristotle.* Hildesheim, Germany: Georg Olms.

Greene, B. (2004). *The fabric of the cosmos: Space, time, and the texture of reality.* New York: Alfred A. Knopf.

Johnson, S. (2002). *Emergence: The connected lives of ants, brains, cities, and software.* New York: Charles Scribner's Sons.

Liu, A. (2004). *The laws of cool: Knowledge, work, and the culture of information.* Chicago: University of Chicago Press.

Longino, H. (2001). *The fate of knowledge.* Princeton, NJ: Princeton University Press.

Lupia, A., McCubbins, M., Popkin, S., Kulinsky, J., & Chong, D. (Eds.). (2000). *Elements of reason: Cognition, choice, and the bounds of rationality.* New York: Cambridge University Press.

McCarthy, J. (1988). Epistemological challenges for connectionism. *Behavioral and Brain Sciences, 11*(1), 21-48.

Moser, P., & Van der Nat, A. (2002). *Human knowledge: Classical and contemporary approaches.* New York: Oxford University Press.

Nozick, R. (1994). *The nature of rationality.* Princeton, NJ: Princeton University Press.

Ophir, A., & Shapin, S. (1991). The place of knowledge: A methodological survey. *Science in Context, 4*(1), 3-21.

Pesic, P. (2004). *Abel's proof: An essay on the sources and meaning of mathematical unsolvability.* Cambridge, MA: MIT Press.

Pinkas, G. (1995). Reasoning, nonmonotonicity and learning in connectionist networks that capture propositional knowledge. *Artificial Intelligence, 77*(2), 203-247.

Polanyi, M. (1966). *The tacit dimension.* New York: Anchor Day Books.

Ridley, M. (2001). *The cooperative gene: How Mendel's demon explains the evolution of complex beings.* New York: The Free Press.

Rubinstein, A. (1997). *Modeling bounded rationality.* Cambridge, MA: MIT Press.

Wasserman, S., Faust, K., & Granovetter, M. (Eds.). (1994). *Social network analysis: Methods and applications.* New York: Cambridge University Press.

Watts, D. (2003). *Small worlds: The dynamics of networks between order and randomness.* Princeton, NJ: Princeton University Press.

Ziolkowski, T. (2000). *The sin of knowledge.* Princeton, NJ: Princeton University Press.

Endnotes

[1]   The reader may wish to consult, for example: Close (2000), Ridley (2001), Glanvill and Fabian (1979), Longino (2001), Pesic (2004), and Moser and Van der Nat (2002).

[2]   See the insightful description of Einstein's discoveries in Greene (2004).

[3]   The reader may wish to consult: Ziolkowski (2000), Agamben and Attell (2004), Capra (1997), Ophir and Shapin (1991), Collins (1990), Polanyi (1966), Lupia, McCubbins, Popkin, Kulinsky, and Chong (2000), and Galison (1999).

[4]   For instance, the paradoxes of knowability, zero knowledge, and complete knowledge provide a platform for further explorations of such concepts as the beginning of life and the existence of a supreme being. By illuminating instances of extreme conditions (such as zero or total knowledge), paradoxes contribute to the extension of our thinking, forcing us to venture beyond accepted norms and traditional thinking.

[5]   Readers interested in Fitch's paradox may consult, for example, the following sources: Fitch, F. (1963). A logical analysis of some value concepts. *Journal of Symbolic Logic, 28*(2), 135-142; Ejerthed, E., & Lindstrom, S. (Eds.). (1997). *Logic, action, and cognition: Essays in philosophical logic.* Boston: Kluwer Academic; Williamson, T. (2000). *Knowledge and its limits.* New York: Oxford University Press.

[6]   There is a problem in continuous clustering of knowledge that may be similar to Karl Marx's hypothesis that in a purely capitalistic society, without controls and regulations, one person may accumulate all available wealth. Similarly, continuous clustering may accumulate all available knowledge. How complex and conceptual can we get? I am arguing here that human limitations of attaining and clustering sensorial inputs, combined with the limitations of clustering KANEs and creating complex knowledge architectures, all contribute to the relatively poor performance of humans in generating and accumulating knowledge. For

instance, only in the past few centuries has there been some distinguished progress in generating knowledge about our universe, whereas in the many thousands of years of human existence, the effort to generate such knowledge has been almost nonexistent.

[7]   I am addressing these philosophical issues with the aim of a brief discussion that would not ignore these problems, and with the hope that the reader will bear with me in this endeavor.

[8]   A more detailed description is given in Part III.

[9]   The reader may wish to consult: Davidson (2004), Gigerenzer and Selten (2002), Nozick (1994), Rubinstein (1997), Elster (1989), Liu (2004), and Foss, N. (2001). Bounded rationality in the economics of organization: Present use and some future possibilities. *Journal of Management and Governance, 5*(3-4), 401-425.

[10]   To evoke a phrase attributed to baseball great Yogi Berra: "It's deja vu all over again."

11.   The reader familiar with artificial intelligence will notice that the term "continuous clustering" is used to denote a method for visualization of a field of data. Some illustrative publications include: Garcke, Preuser, Rumpf et al. (2001), Pinkas (1995), Boutsinas (2004), and McCarthy (1988). The basic idea is the same: repeated clusterings of elements with the purpose of creating a framework or an entity that will better describe and perhaps even explain the disposition and the behavior of these elements when they are conjoined.

[12]   The size of the amygdala in the brain was found to be smaller in cocaine addicts, which is an indication of the role that the size and location of areas of the brain may have on the performance of certain functions. See, for example, Marks, M. et al. (2004). Decreased absolute amygdala volume in cocaine addicts. *Neuron, 44*(4), 729-740. Also see Knierim, J. (2003). Hippocampus and memory: Can we have our place and feat it too? *Neuron, 37*(3), 372-374. The author suggested that "hippocampal cells respond to the combination of spatial location and conditional stimuli in a nonspatial fear conditioning task."

[13]   The reader may consult Buchanan (2003), Watts (2003), Girolam (1999), and Domhoff (2002).

[14]   A valid question would be: Do criteria for clustering change as complexity increases? Regardless of whether we answer in the affirmative, the criteria would be modified to accommodate complexity, but would

not violate the principle of continuous clustering and its implications for the structure of knowledge.

[15]   I am cognizant of the fact that I may be oversimplifying a complex area of research. Studies into neural networks have generated many insights into mapping self-organizing entities and transfer mechanisms among different components (e.g., Johnson, 2002; Wasserman, Faust, & Granovetter, 1994). Nevertheless, I believe that the differences between the structure of knowledge and the structure and function of neural networks are irreconcilable. I am convinced that for our understanding of the structure of human knowledge, neural networks are of little help. See Chen, C., & Paul, R. (2001). Visualizing a knowledge domain's intellectual structure. *Computer, 34*(3), 65-71; Lehrer, K. (2000). *Theory of knowledge.* Boulder, CO: Westview Press.

[16]   We now know from MRI mapping that dreams appear to occur in the same location in the brain also used to identify images, such as visual memory and facial recognition. There is no need for the brain to divert effort to other regions. If dreams are some attempt at comparison, such "testing" of knowledge will occur in the same locality in which it was created originally.

[17]   I would add that the "Turing test" (discussed earlier) is irrelevant. It is based on the notion of artificial computing, not on knowledge generation. There is a vast literature on the meaning of the test. Alan Turing (1912-1954) was concerned with: "Can machines think?" Critics had argued that the test is merely a test for human intelligence. Simply put, if a machine can "imitate" human thinking and intelligence, it would pass the test. I argue that, regardless of how ably a machine can fool a human, one cannot conclude that it generates knowledge as humans do. I will return to this topic in Part III of this book, in the discussion of how knowledge is stored in machines and diffused by them.

Section II

The Progress of Knowledge

Chapter VIII

The Evolutionary Model:
Selection and Progress of
Knowledge

The Knowledge Society

It is often heralded that we live today in the "knowledge economy" and the "knowledge society." It has become quite a traditional endeavor to divide the history of human progress into three main phases: the agrarian economy, the industrial economy, and the contemporary economy. The agrarian economy was based on labor and land, the industrial economy on labor and capital, and the contemporary economy on labor and knowledge.

But this is only part of the story. The agrarian society depended for its survival not only on the land in which crops were grown and animals husbanded. To be successful and to avoid sudden periods of famine, the agrarian society depended on a varied, albeit rudimentary, amount of knowledge. Agrarian economies depended on knowledge of climatic changes and a degree of understanding of lunar and solar phases. They depended on some knowledge of mathematics, geometry, architecture, and engineering to build irrigation canals, roads for the transport of foodstuff, warehouses to store food surpluses, and an understanding of the patterns of the flooding of the rivers on the shores of which they built their villages and cities. They depended on knowledge of geology, zoology, and botany to understand how to plant and care for seeds, how to rejuvenate and irrigate the soil, how to husband animals, and how to use plants and animal products for habitation, clothing, defense, and other functions of a primitive society.

The Biblical story of Joseph, son of Jacob who translated Pharaoh's dream of the seven lean years, is an illustration of the need for knowledge of the cycles of the flooding of the river Nile, and the knowledge necessary for planning, harvesting, and storing excess foodstuff for the upcoming "lean years" to avoid massive hunger.

The industrial revolution introduced the role of capital and the decline in the importance of land. Factories could be built on limited tracts, even within city boundaries. New types of knowledge were now required to run these factories, to obtain and transport raw materials, to finance these ventures, and to bring the finished products to the consumer in the marketplace. The industrial economy was dependent upon a set of constraints: limited capital, raw materials, and skilled labor. There were also constraints of competition, risks to production and to commerce, and the intricacies of the political and social environments within which industries had to function. All of these brought about the decline in the need for knowledge of nature and the rise in the need for knowledge about the economic, social, and production dimensions of the emerging social and economic arrangements of the industrial age.

The constraints and demands of the industrial age were also the driving force behind the rise in the diffusion of such diverse types of complex knowledge. The industrial age produced massive public and privately funded education and training of the new cadres of skilled labor and the producers of the increasingly complex knowledge necessary for the growth and maintenance of these industries.

So, what is different about the contemporary "knowledge society" and the "knowledge economy"? Six factors constitute the variables that explain the increasingly cardinal role that knowledge plays in our world today:

1. The sheer *volume of knowledge*—both technical and other—being produced and exchanged is almost beyond grasp. The rate of production of human knowledge is breathtakingly accelerating on a daily basis.

2. We have developed *tools* and *instruments* to process and utilize knowledge which are increasingly powerful and sophisticated. We can nowadays manipulate any desired quantities of knowledge, for almost any needs.[2]

3. We have a much larger sector or segment of *people who possess* knowledge (technical, managerial, organizational, and industry and job-specific), who are in pursuit of it, and who utilize it as the principal tool of their profession. Moreover, they depend on ever-increasing arsenals of knowledge for their success and their survival.

4. Knowledge *has become specialized* and even further down to sub-specialties. This phenomenon of knowledge becoming increasingly specific has resulted in the creation in society and the economy of "silos," in which specialists strive for even more specialization and the enhanced idiosyncratic nature of their knowledge.

5. Knowledge is more and more *embedded in products and services* we normally consume and in the jobs we practice. A generation ago such knowledge was embedded only in highly technical and very expensive equipment and systems. Nowadays, mundane and household products are embedded with knowledge previously confined to exclusive and larger-scale apparatuses, mostly in the defense and intelligence areas.

6. Knowledge has become *objective, worldwide,* and to a large extent, a *commodity.* It can be easily stored, transformed, learned, imitated, and utilized. Since knowledge is no longer the domain of the few, it is transforming the ways in which certain professions and certain processes are conducted.[3]

Our parents were satisfied with a high school education to hold a lifetime employment in factories or in services. Today, a college degree is a necessity, as the knowledge thus acquired has turned into a commodity. Since knowledge increases incessantly, the need for continual updating of one's knowledge base leads to obsolescence of the existing pool of knowledge we possess at our place of employment.

Furthermore, because knowledge is a worldwide phenomenon (as a truly global "equalizer"), outsourcing of jobs has recently been upgraded to more "knowledge-based" skills needed for even sophisticated employees. Under the overall umbrella of "globalization," geography ceased to be a barrier to the transfer of knowledge and to the outsourcing of knowledge-based jobs.[4]

With this scenario as the backdrop of our contemporary life, it is paramount that we gain a profound understanding of how we generate knowledge and how it progresses. This is no longer the pastime of some curious intellectuals and philosophers. This is becoming the mainstay dimension of our existence in all parts of the world—the developed economies and the less-developed countries. Knowledge is ubiquitous and highly disrespectful of political and economic frontiers. Although in the developing world only small segments of the population is versed in the new knowledge economy, their impact on global economic trends is substantial and will ultimately lead to the distribution of knowledge in wider and more diverse segments of their populations.[5]

Models of Progress: The Power of Evolution

How knowledge progresses can perhaps be modeled by two major approaches: *evolution* and *cumulation*. The reasons for choosing either model will be discussed later in this chapter. The issue before us is the mode in which knowledge (as defined in Part I of this book) advances and increases in size and perhaps in quality or other attributes. However we choose to go, it is clear from the outset that there is a context by which the "stock of knowledge" increases or grows, and that knowledge cannot be restricted to a "piece" of knowledge or a "nugget" or whatever term we had applied earlier in the book. Knowledge only makes any sense when viewed as a "pool" or a "stock" in which multiple items, nuggets, or pieces are grouped.

We considered earlier in the book how such elements of knowledge are generated and how they are structured, grouped, and assembled into select architectures. But, how does knowledge grow, progress, and advance to form "bodies" of knowledge with their distinct uses and functions for the knower if, as I suggested, they are restricted to the "knower"?

The model of evolution has been a powerful approach to the description of how knowledge increases and progresses. The theory of evolution is an elegant theory with a strong power of explanation of the diversity of biological forms. Traditionally, students of knowledge had been mostly concerned with the truth of knowledge and its justification in models of ethics and utility.[6] But, as soon as the theory of evolution took hold for biology, it was a short path to attempts by philosophers and social scientists to apply it to the progress of knowledge. It made a lot of sense if one starts with the assumption that biological organisms cannot survive and adapt to their changing environments without knowledge about such environments.

Early attempts are embedded in Sir Karl Popper's epistemology of scientific inquiry (see Popper, 1962, 1972, 1995). He argued that scientific theories must be subjected to tests of refutability, and that all scientific theories must be falsifiable.[7] If they cannot withstand the rigorous test of empirical verification, theories must be rejected. Hence, this process is very similar to natural selection in biological evolution.

Evolutionary Epistemology

Initially introduced by Donald Campbell (1987, 1994) to describe a hierarchical view of the link between biology and knowledge of generation and its progression, Campbell suggested that knowledge progresses by a process of evolution, from genetic transformations all the way to scientific discoveries. This hierarchy had eleven steps: (1) genetic adaptation, (2) non-mnemonic problem-solving, (3) vicarious locomotive devices, (4) instinct, (5) habit, (6) visually supported thought, (7) mnemonically supported thought, (8) socially vicarious explorations, (9) language, (10) cultural accumulation, and (11) science.[8] This model was based on two key principles. First is that in the beginning knowledge is generated in a "blind" mode, at which stage there is a selection and retention of "good" knowledge and the discarding of "bad" knowledge. This is the principle of *blind-variation and selective retention.*

Second is the principle of *vicarious selection* by which knowledge that has been already retained (after initial selection) serves as a benchmark for comparison with new knowledge. Thus, selection is not blind and new knowledge can be selectively accepted or rejected with reference to an existing pool of knowledge that serves as criteria for "fitness."

The selection and retention processes are embedded in the eleven "nested hierarchies." As additional knowledge is selected and retained—while ascending the hierarchy—organisms with such knowledge become more adaptable to their environments. Knowledge progresses through successive selection processes, where the "fit" knowledge is retained and the "unfit" is rejected. All this is in support of the organism generating and using this knowledge as a tool in its evolutionary pathway and its struggle for survival in a changing and competitive environment.

This view of knowledge, from the genetic adaptation of organisms to their use of science, assumes that knowledge is an instrument in the struggle for survival. In the beginning there are genetic activities, where knowledge is embedded in the genetic material. As organisms (including humans) begin to transact with the external world, they require inputs of knowledge about this world in order to survive. The more complex and sophisticated these inputs and their selection processes, the more the organism is able to survive and adapt to the competitive and the unforgiving environment. Thus, the last nested hierarchy—*science*—has the most sophisticated procedure for the selection, retention, and rejection of knowledge.

Evolutionary epistemology therefore suggests that knowledge progresses in a consistent mode toward more complex selection and utilization of knowledge by organisms in their environments. The more they generate and use knowledge, and the higher is the state of selection, the more these organisms secure intelligence, and the more they possess improved instruments to compete in hostile surroundings.[9]

Why Knowledge is not Evolutionary

The appeal of the evolutionary model is mainly in the simplicity and elegance of its tenets in explaining diversity and change. In previous publications I suggested at least seven reasons (Geisler, 2001).

Biology vs. Knowledge

Evolution and its tenets of natural selection and survival of the fittest are good explanations for the way biological entities survive, diversify, change, and perish. Knowledge is not a biological entity. In the leap from the "Amoeba to Einstein" and from genetic transformations to science, evolutionary epistemologists have failed to distinguish between the *process* or *method* of scientific discovery and the development of knowledge. However much we reify knowledge and provide it with form, it will not and cannot emulate biological organisms. Knowledge may be a tool with which biological organisms conduct their battle for survival, but it does not follow that the *tool* behaves in a way similar to its *user.*

Extinction of Biological Species

Natural selection and survival of the fittest means that those biological species who cannot survive will perish. Once the species is extinct, it is gone forever. Knowledge, however, does not entirely disappear. Theories which were in use and discredited become theories-in-disuse. Furthermore, the discredited theories are used as a learning device to prove the viability of competing theories. Nature does not use extinct species as a mechanism to blend with current species. In fact, genetic material from rejected elements of the species or an entire species disappear from the genetic pool of the surviving organisms. This is the name of the game of biological evolution. Ideas such as the "phlogiston" or "perpetual motion" do not disappear just because they had been proven to be unworkable.[10]

Process vs. Method

As we ascend Campbell's hierarchy to the stage of scientific inquiry, it becomes abundantly clear that evolutionary epistemologists confuse the *method* of scientific inquiry with the *process* by which knowledge progresses. The scientific revolution, which gave us the *scientific method* of acceptance or rejection of hypotheses and theories, was about making the generation of knowledge more objective. The aim was to create an objective procedure, away from religious or political influences and dogma, and in a mode subjected to experimentation and easily exchanged among scientists, regardless of their nationality, religion, language, and cultural constraints.

But this was the *scientific method,* not the *process* by which knowledge progresses. The method dictates which theories we accept, yet does not describe nor explain how knowledge progresses, grows, and accumulates. Imagine the following allegory. In describing a basketball game, the sports television analyst relates to his audience the techniques by which players move about the court, starting with the strategy each team uses in their attack and defense all the way to the decisions whether to pass the ball or throw it at the basket. What the analyst fails to do is describe the *game,* how it progresses, and how it ends. The scientific method is about how we move around the court. The progress of knowledge is the process by which teams play the game: how the points add up in the score of who wins and who loses.

Need for Diversity

Diversity is a fundamental requisite for biological evolution. The idea is that large populations undergo processes of evolutionary changes. But knowledge does not require a large pool from which to draw those theories or hypotheses that will or will not survive. The body of knowledge at any point in its progress is not necessarily large, in fact it may be quite limited. The competitive pool in knowledge may consist of only two competing hypotheses.

Genesis and Conjugation

Human knowledge is generated as a result of our interaction with the environment, and by absorbing sensorial inputs from it. However we process such inputs, they end up representing environmental events or scenes. Unlike biological beings, unique environmental events or scenes are not members of a species. We may classify and categorize them to our contentment, but they lack inherent traits that group them as a distinct species. Hull (1988) argued that scientific knowledge is to scientific disciplines as biological entities are to biological species. However, knowledge about the environment is "conjugated" to the event or scene it represents and cannot be treated independently of the event.

Higher-Order Constructs

According to evolutionary epistemology, knowledge accumulates by a process by which complexity increases as criteria for selection are more advanced and theories are rejected with increased sophistication. If knowledge is formed by clustering, natural selection is not an option in generating higher-order constructs. For these constructs to appear, there is a need for an existing body of accepted knowledge on which to build further. The issue, then, is not criteria for refutation but criteria for additiveness to weave a larger picture. This is another aspect of the confusion between method and process.

Boundaries, Expansion, and Equilibrium[11]

Biological evolution is predicated on the limits that competitive environments impose upon the species that inhabit them. There are boundaries to growth and expansion. The evolutionary process is viewed as reaching some form of equilibrium, such as Gould's notion of "punctuated equilibria." This means that changes, mutations, and the pre-determined conditions of the natural environment dictate limits to how much growth is possible and how fast it can be achieved.

Knowledge, on the other hand, is not bounded by these constraints to the extent that its users are so limited. The growth of knowledge depends upon its user's abilities to collect inputs from its environment and to process them. But there are no pre-set limits to this effort.[12] In relative terms, knowledge can expand not only within the individual's capacity to absorb and process, but also in a multiplier effect of a population of individuals. Finally, knowledge does not require periods of equilibrium. It progresses continually, without the need for a balance in its structure or process.[13]

The Problem with "Fitness" and the Analysis of Functionality

Unlike biological entities, knowledge progresses without the constraint of "fitness." In the case of biology, natural selection and the survival of the "fittest" lead to the extinction of some species and the survival of others. The criterion of "fitness" refers to the entity's ability to adapt to its environment, hence to "fit" its changing demands.

Knowledge can be viewed as an instrument leading to better adaptation, hence to survival. This would be the function of knowledge in the struggle of biological entities for survival. But, knowledge itself does not need to be "fitted" to this function. Fitness is not a quality or attribute that is inherent in a given item or piece of knowledge. Rather, such attributes would be relative to the needs of the biological entity (the "knower") at a given situation and with a specific need for survival.

When biological entities survive by the process of natural selection, they are better fitted to adapt to their environment. The emphasis here is on "better." They are better equipped to confront environmental pressures and have a higher probability of survival. Such "betterment" is not a characteristic of knowledge as it progresses, nor a criterion for its selection. Even in the Popperian approach, theories are refuted or provisionally accepted—not improved upon for later acceptance.[14]

Knowledge is not Evolutionary

The progress of human knowledge cannot be explained by a model of evolutionary advancement, by natural selection, or by the survival of the fittest. There is a need for a different model, one that would be tailored to the way in which knowledge is structured and to its inherent qualities.

In the reasons I listed why knowledge does not evolve as do biological entities, one of the key notions was the reification of knowledge. As I discussed in previous chapters, we tend to consider "knowledge" an entity, thus giving it attributes about form and movement. Knowledge may be viewed in a reified manner when we consider a "body" of knowledge, which is the cumulative assembly of units of knowledge. This means that such a "body" is now treated as an entity and can be seen as performing a function for the "knower" who has possession of such a body. Regardless of whether we consider progress of the elements *within* this body or the progress of the *entire* body as one unit, the criteria for growth, acceptance, or rejection would be subjective, so that some bodies of knowledge may be useful to knower A and of no use to knower B.[15]

By virtue of it being a "body" of knowledge, the attributes of this body are in line with the accumulation of knowledge that make up this form. Therefore, as soon as we consider knowledge as a congregation of elements or components, we are introducing the notion of cumulation and additiveness. This is a key ingredient in the model of cumulation that I discuss in the next chapter.

References

Baum, J., & McKelvey, B. (Eds.). (1999). *Variations in organization science: In honor of Donald T. Campbell.* Thousand Oaks, CA: Sage.

Campbell, D.T. (1987). Evolutionary epistemology. In G. Radnitzky & W. Bartley (Eds.), *Evolutionary epistemology, rationality, and the sociology of knowledge.* LaSalle, IL: Open Court.

Campbell, D.T. (1994). How individual and face-to-face group selection undermine firm's selection in organizational evolution. In J. Baum & J. Singh (Eds.), *Evolutionary dynamics of organizations* (pp. 23-28). New York: Oxford University Press.

Campbell, D.T. (1996). Unresolved issues in measurement validity: An autobiographical overview. *Psychological Assessment, 8*(4), 363-368.

Cohen, L., & Young, A. (2005). *Multisourcing: Moving beyond outsourcing to achieve growth and agility.* Boston: Harvard Business School.

Geisler, E. (2001). Good-bye Dodo bird (Raphus cucullatus): Why social knowledge is cumulative, expansive, and nonevolutionary. *Journal of Management Inquiry, 10*(1), 5-15.

Hull, D. (1988). *Science as a process: An evolutionary account of the social and conceptual development of science.* Chicago: University of Chicago Press.

Popper, K. (1962). *Conjectures and refutations.* New York: Basic Books.

Popper, K. (1972). *Objective knowledge: An evolutionary approach.* Oxford, UK: Oxford University Press.

Popper, K. (1995). *Objective knowledge: An evolutionary approach* (revised ed.). Oxford: Clarendon Press.

Vashishta, A. (2006). *The offshore nation: Strategies for success in global outsourcing and offshoring.* New York: McGraw-Hill.

Endnotes

[1]   The term *progress* is used throughout this book in a generic sense. It does not necessarily mean "improvement" or "evolution" due to the generation of knowledge, nor does it necessarily mean a positive development due to knowledge. Perhaps *growth* would be a better term that describes the way in which knowledge adds up, accumulates, and increases in quantity and quality.

[2]   The Internet is, of course, a good example of such an instrument where knowledge is stored and can be retrieved and utilized for many needs. Other forms of electronic exchange are also included in these tools, for example, teleconferencing, telemedicine, and teleradiology.

[3]   For example, patients can now obtain medical knowledge from the Internet. They feel more "empowered" and in better control of their condition. They tend to use this knowledge in their interaction with their caregivers, thus changing the way medicine is practiced in some areas of healthcare.

⁴   The reader may wish to consult an excellent story about outsourcing by Aaron Bernstein in *Business Week,* December 6, 2004. He reported that by the year 2015, more than 6% of nonfactory jobs will be outsourced by America's largest companies. These jobs will be in such areas as computers, law, architecture, life sciences, and management—that is, where specialized knowledge is the key ingredient in the profession. Also see Vashishta (2006) and Cohen and Young (2005).

⁵   In 2005, the United Arab Emirates (UAE), for example, had established the Dubai Outsourcing Zone (DOZ). The zone is designed to attract businesses from the high-cost environments of developed countries. Economic incentives are offered, such as full ownership for foreign investors and no income taxes. Since the zone is directed at people and knowledge-intensive industries, Dubai is also offering better living conditions and other economic benefits that compete directly with outsourcing giant India.

⁶   See Chapters II and III. Recently, functional brain imaging has shown that lying and truth telling cause activity in different parts of the human brain. By using the functional MRI, researchers concluded that it takes more effort by the brain to tell a lie, and that lying is conducted in those parts of the brain which are also responsible for emotional reactions.

⁷   Clearly, these principles also apply to hypotheses or propositions.

⁸   See Chapters II and III in Part I of this book.

⁹   As discussed in Chapters II and III, in his later papers Campbell (1994, 1996) had recognized the methodological and conceptual consequences that arise from the notion of nested hierarchies. Higher-level organisms evolve by selection criteria that may shift the balance of natural selection. For example, human use of science and technology in our struggle for survival has led to the extinction of species around the globe. I should add that, central to evolutionary epistemology is also the notion that lower-level entities exhibit traits of selfishness, whereas higher-order organisms which congregate in complex societies utilize knowledge to maintain the integrity of these social formations. Campbell called this "clique selfishness."

¹⁰   If knowledge is viewed as a function of the user, then clearly the key criterion for what happens to knowledge is not whether it disappears or becomes extinct, but whether it is in *use* or *disuse* by the user.

[11]   These reasons for why knowledge is not evolutionary have failed to convince scholars who support evolutionary epistemology—see Baum and McKelvey's (1999) edited book in honor of Donald J. Campbell. Personal communications by several authors in this volume had expressed agreement with most of the reasons in this chapter, yet they rejected the totality of my arguments against evolutionary epistemology and my model of cumulation. Nonetheless, my view remains unshakable that knowledge does not progress in an evolutionary mode.

[12]   This comparison is, of course, in relative terms. Obviously there would be limits to the sensorial inputs available to creators of knowledge and to the capacity of their brains to absorb and to process those inputs.

[13]   Readers familiar with Kuhn's paradigm shifts may argue that such punctuated "revolutions" in science do resemble the puncturing of equilibria. However, as I understand Kuhn's description of this phenomenon, the shift is in the paradigm of a given perspective or "school of thought" in science, namely, a social phenomenon rather than modifications or revolutions in the *knowledge* itself. See also my discussion of the problem of "fitness" and functionality analysis of human knowledge. Therefore, Kuhn's ideas do not apply here.

[14]   Evolutionary epistemologists seem to easily confuse the functional worthiness of knowledge for the struggle of biological entities in their environment with the inherent attributes of knowledge and its progress. As I explain in Chapter VIII, knowledge does not "improve" by evolving, rather it becomes more valuable as it grows. In other words, we do not replace "bad" knowledge with "good" knowledge—we simply know *more* about a certain topic, hence we now derive more value out of what we know. The process is therefore not evolutionary; it is cumulative.

[15]   In science, for example, some discarded theories had adequately explained certain physical phenomena until their demise (e.g., ether, which explained the medium for propagation of light).

Chapter IX

The Cumulation Hypothesis and the Model of Continuous Cumulation

Why a Model of Progress of Knowledge is so Important

We generate, store, organize, and utilize knowledge in order to represent complex phenomena in ourselves and in our environment. We accumulate knowledge consistently and furiously from the moment of our birth to the last breath of air we take. Further, we catalog and preserve knowledge in various forms: in our memories, in oral story telling, and in physical modes such as written language, architecture, art, and with our genetic materials. Thus, we impart knowledge through our own biological traits and by all the means we use to impact our environment.

The growth and progress of such knowledge is paramount to our understanding of how knowledge can be used, transferred, shared, and diffused.

The transfer of knowledge among people and across cultures and time has been the realm of research by sociologists, philologists, and information and linguistic scientists.[1]

In effect, there are three key reasons why a model of progress of knowledge is so important. First, the *philosophical* topic of how knowledge grows and expands has been a vexing problem for generations of philosophers. Second is the *sociological* issue of how knowledge grows and how it diffuses within social systems.[2] Thirdly, it is of paramount importance for the *design and use of databases, knowledge systems,* and *knowledge warehouses.* A good model of the growth of knowledge will allow us to understand how to better store, maintain, and access knowledge in warehouses specifically designed to accommodate the ways in which the stock of knowledge increases.[3]

If we accept the evolutionary model, we would have to design warehouses that are classified by the nested hierarchies as proposed by evolutionary epistemologists. Acceptance of any item of new knowledge into the warehouse would be subject to refutation. Furthermore, as knowledge accumulates in the warehouse, maintenance of it must take into account the potential utilization of such knowledge, and would also take into account its obsolescence and its replacement by newly accepted knowledge in the course of the natural selection process.

If, however, we accept the model of cumulation I am proposing here, the design, maintenance, and use of knowledge warehousing will be very different and, in my view, better suited for use by social, economic, and organizational entities. So, what is the model of cumulation and why is it more advantageous than the evolutionary model?

The Cumulation Hypothesis: The Unity in Modeling

The unity of Part I (the structure of knowledge) and Part II (the progress of knowledge) rests on the sharing of principles between the model of continuous clustering of sensorial inputs and the model of cumulation. The principle of how knowledge is structured and progresses in both models rests on the underlying notion that the generation and progress of knowledge is conducted by a device which we call the "brain," and that it starts as a "tabula rasa," a clean slate, into which knowledge is deposited in a cumulative manner.

The brain is considered a processor of knowledge. It is viewed as a factory, a manufacturing plant in which sensorial inputs are the raw material transformed into knowledge by ever-recurring clustering. In fact, the recent flurry of publications on the link between genetics and brain structure has highlighted general interest in this topic.[4] Since the discovery that very few genes are responsible for the formation of the human brain, many cognitive psychologists and other scholars have turned their attention to the link between the *physical* brain and its function as a *mind,* a *consciousness,* and a generator of knowledge.[5]

In my view, the genes we inherit are only responsible for the structure and the physical nature of this factory, the brain. We inherit a blueprint for a mechanism that is capable of receiving sensorial inputs and transforming them into knowledge. This blueprint provides a "turnkey" operation, in which we receive a factory whose design and inherent effectiveness or competency is the product of the gifts of our species through evolution, and the gifts from our parents and their progenitors. Any defects in this factory may be attributed to the original blueprint and to its development—as part of our development from birth to adulthood.

Imagine a construction company that designs and builds a "smart" manufacturing plant—to be handed to its owners and operators. This is a "turnkey" project. Clearly, in the design and construction of this plant, the construction company has employed all the knowledge it acquired in many disciplines, such as building, floor layout, energy inputs and usage, and the installation and commissioning of machinery, software, and communication technologies. This is how our genes band together to form a functioning brain.

The allegory of the "turnkey plant" may explain the need for continuous flow of sensorial inputs and what happens when we suffer from *sensorial deficiency.* For example, when one of the senses is diminished or ceases to function, there is a compensated effort in the brain. In essence, blind people do not possess better hearing (they are not designed for it), but the brain makes better use of auditory inputs, just as a functioning manufacturing plant will make better use of its existing raw materials when a shortage occurs in one of the materials needed for production.[6]

Embracing the notion of the brain as a turnkey operation which later would be developed in childhood may explain recent findings from studies of children and their social development. Stanley Greenspan, professor of psychiatry at George Washington University, has argued that watching television does not help to develop social skills in very young children and may, in fact, hinder such development.[7]

Viewed from the perspective of the structure and progress of knowledge, children to age 4 who excessively watch television are exposed to limited sensorial inputs, whereas children who interact with other children receive a richer variety of inputs. Thus, the clustering of only visual and auditory inputs results in a less enriched environment from which the brain can cluster, whereas social interactions offer additional sensorial inputs for clustering.[8]

The *cumulation hypothesis* postulates that knowledge progresses as incremental additions to a platform in an ongoing array of incremental configurations. Living, rational organisms are continually adding to their stock of knowledge. Knowledge is continually generated from the clustering of sensorial inputs and added to a growing and ever-expanding body of what we know as individual knowers.

The same principle also applies to the cumulative body of knowledge of a collective of individuals or society, and over time. When humans grasped a way to preserve the knowledge they possessed so it would continue to exist after their death, the cumulative body of individual knowledge became the body of knowledge of the social entity. Language, artistic expression, and writing allowed individuals to contribute to the public stock of knowledge.

Yet, knowledge is embedded in the mind as the result of clustering of sensorial inputs, so much of it cannot be transferred to others. There is a need for a *shared platform,* such as language and the grasping of concepts, to have some form of exchange or knowledge.

Imagine a scenario where one individual possesses an extraordinary amount of knowledge, whereas the remaining people possess very little knowledge. Two outcomes are thus possible. First, few tasks can be accomplished because they would depend on the knowledgeable person who would find it almost impossible to transfer what he can transfer to others, due to the fact that their puny amount of knowledge would lack the shared platform for meaningful and effective exchange.

A corollary scenario would be a number of people with some or much knowledge, but none having an exceeding amount. Tasks would be easier to accomplish, since exchange will be more viable. In both scenarios, cumulation is vital. The more knowledge is gained by as many people as possible, the higher the probability that some of this knowledge will be exchanged and transferred to the benefit of the entire group.

Clustering and *cumulation* are two expressions of a single phenomenon by which the mind generates knowledge and allows for its progress and growth. The mind works in a cumulative mode so that the stock of knowledge thus generated can be accessed in a highly economic and efficient manner. How does this work? We utilize very few items of knowledge in the existing stock to form complex representations of physical phenomena in our environment. The larger our accumulated stock of knowledge ("knowledge base"), the fewer items or KANEs are needed to frame a construct. The larger the stock of knowledge, the easier it is to make comparisons with a newly clustered set of knowledge, and the more efficient our mode of understanding our environment.

In the competing model of evolutionary progress, there is a dependency on external forces that impact the process of survival of the fittest and reliance on selection from within a population of potential knowledge. This process does not permit the kind of efficiencies in the operation of the mind that are offered by the cumulation hypothesis.

The Cumulation Hypothesis: The Allegory of the Wall of Knowledge

Imagine the progress of knowledge as the erection of a brick wall.[9] The bricks used in the construction are lying around. The bricklayer in this case is the mind. Following a given plan or design, the bricklayer adds the bricks, one by one, in rows of a predetermined layout. He would select bricks of different shapes, designs, and colors. As he lays them in row after row, a pattern emerges. This is the specific architecture of the wall. There are many possible architectures and patterns, and as the wall grows, there may be different patterns in different parts of the wall.

Knowledge progresses in the way a wall is constructed. The body of personal knowledge resembles a brick wall, painstakingly assembled from a variety of different bricks held together by what we would define as "memory" in the human brain. As we add items of knowledge—as we do with added bricks—the addition depends on the existing architecture. Each additional knowledge rests on the previously accumulated body.

Sir Isaac Newton (1643-1727) once commented that he "stood on the shoulders of giants." Stephen Hawking chose this phrase as the title of his 2002 book on the great works in physics and astronomy. This statement, in my view, summarizes and celebrates the cumulation hypothesis: that knowledge is additive and that each addition rests on the existing body of accumulated knowledge.

The Model of Continuous Cumulation

Knowledge progresses by continuous cumulation and expansion. There are four attributes to this model.[10] The first is *sedimentarity.* Knowledge is added in a mode that resembles geological formations that accumulate by sediments. This is the bedrock of what we would call "prior art," in that every layer that is added to the sediment rests on all the previous layers.

The second attribute is the existence of *systemic flexible boundaries.* Human knowledge is expansive—it grows and expands in continuous cumulation. I entertain here the notion that the more knowledge accumulates and the more its richness and diversity, the more flexible will be the boundaries for its expansion. This means that human knowledge has few limitations in how much it can grow and expand. Unlike biological organisms which are severely limited by their genetic make up, human knowledge is, for all practical purposes, without limits, except those imposed by the capacity of the brain to process knowledge and the capacity of the warehousing we construct for storing our collective knowledge.

A third attribute of the progress of knowledge may be compared to the "expanding universe." As the richness of the pool of knowledge increases, the accumulation of any and all additional sediments of knowledge will be facilitated. This means that when the wall of knowledge is large enough, there are more "opportunities" for additional cumulation, and there are added possibilities for creating different architectures. The more raw material exists in the body of knowledge, the more interactions are possible, hence the growth will be, for all practical reasons, "ever expanding."

Finally, the attribute of *complementarity* suggests that as knowledge accumulates, there are synergetic benefits that accrue to the body of knowledge from the diversity in the pool.[11] Complementarity is a phenomenon in which the benefits accrued from cross-fertilization of added knowledge will open up additional ways and means for knowledge to join the pool. Imagine the interaction of neurons in the brain. Complementarity in the interaction of neurons with other neurons creates added "ports" in which new knowledge may be absorbed and processed.

The Notion of Architecture and Continuous Cumulation

The additiveness of items of knowledge by continuous cumulation creates arrangements that we may denote as "architectures" of knowledge. As knowledge accumulates within a given architecture, there is also an increase in the complexity of the pool of knowledge. This occurs because there is an increase in the number of possible arrangements by which knowledge may be organized and new knowledge added to the pool. Another reason is the increase in the diversity or types of knowledge in the pool, leading to new configurations.

For example, consider the pool of knowledge about the structure and behavior of the human cell. As this specific knowledge increases, multiple disciplines are incorporated into the pool: microbiology, flow models, genetics, biochemistry, and bioengineering. These additions of knowledge will be amenable to a variety of possible architectures and so increase our potential understanding of how human cells are structured and the process that they undergo in the human body.

Architectures are a glorified term for arrangements of knowledge, but they also denote an inherent logic in the design and progress of these configurations. Their diversity offers alternative scenarios of how the existing pool of knowledge may describe and explain certain phenomena.

Moreover, when we add layers to an existing architecture (see the allegory of the brick wall), we may be adding complexity. The relationship between added complexity in the arrangement and the progress of knowledge is in the possibility of *reconfiguration* of the existing knowledge pool. This may mean the emergence of a new architecture. Imagine the brick wall. When the architect of the wall wishes to change the shape and pattern of a new section of the wall, he may select a different type of brick, and may even *dismantle* certain parts of the wall and *rebuild* it according to the new pattern.[12]

The Measurement of Cumulation

In the model of continuous cumulation, knowledge is continually added to the stock of knowledge stored in the human brain. If we wish to measure this cumulation effort, we may be searching for measures or metrics of additiveness. This will be a measure of, for example, the units of knowledge gained per amount of effort expended, per time, or per problem or topic for which we collected such knowledge.

Additiveness is a phenomenon well researched in other fields such as information processing and artificial intelligence.[13] In the case of knowledge, additivity would be a measure of the *benefits* accrued from an added unit of knowledge, rather than a measure of the capacity of the container of knowledge or the channels used to propagate it.[14] Uren, Shum, Mancini, and Li (2004) offered a good taxonomy that could begin to classify these benefits of added knowledge:

1. General (improves an existing pool, enables)
2. Problem related (addresses, solves)
3. Supports or challenges (proves, refutes, is evidence for or against, agrees or disagrees)
4. Causal (predicts, envisages, helps to identify causality)
5. Similarity (identical or similar, analogous, or not)
6. Taxonomic (part of, example of, sub-class of)

These benefits or properties of added knowledge provide a qualitative measure of what such knowledge can provide. For the individual knower, each added item of knowledge is compared with the existing catalog of accumulated knowledge, and its contribution is assessed. But, although we have models in information theory and artificial intelligence which can be used to implement such measurement, it is an experimental impossibility to invade the human brain and make such evaluation at the level of individual items of knowledge and their contributions to the existing pool of accumulated knowledge.

Measuring Process with Brain Waves

The measurement of cumulation of knowledge could be done by assessing *outcomes* of additional cumulation (such as the contributions described above) and *processes* (how additional knowledge is processed by the brain).[15] Our current state of technical achievement allows for an indirect measurement, not of cumulation but of brain activity, which indicates certain uses of the knowledge accumulated and processed by the brain. This is done by measuring brain waves.

There are four types of brain waves, classified by their frequency band. Brain waves are electrical signals exchanged among brain cells (neurons) that can be recorded, thus indicating an activity within the brain.[16] The frequency bands (Hz = Hertz or cycles per second) have been associated with certain types of activities or functions.

Delta waves have a frequency of 1.5-4 cycles and are associated with dreamless sleep. Next are *theta* waves (5-8 cycles per second), associated with a state of sleep with active dreaming, usually accompanied by rapid eye movement (REM). *Alpha* waves (ranging between 9-14 cycles) are indicative of relaxation, reflection, and meditation. Finally, *beta* waves (15-40 cycles per second) are present when the brain is very active in a function that requires quick thinking and processing of knowledge.

The classification of brain waves and their relation to the various states of the brain clearly show that it takes the fastest frequency of brain waves to maintain a state of arousal and processing of knowledge. This type of *beta* brain wave is associated with cumulation of knowledge and its processing—although of course we are as yet unable to unequivocally relate the brain's electrical activity with its specific internal functioning at the neuronal level.[17]

Advantages of Continuous Cumulation

Adding layers of knowledge to the existing architecture is a process of continuous cumulation and progressive addition. The advantages of this model to databases and knowledge warehousing are discussed in the next two chapters of this book. Here I emphasize the unique advantage of continuous cumulation for the unity of structure and progress of knowledge. As we generate KANEs through clustering of sensorial inputs, we accumulate them in more complex forms in our memory. Presently we have identified and classified four types of memory: procedural, declarative, working, and external.

Procedural memory is the storage in the brain that contains knowledge about how to proceed in a given action or activity, such as riding a bicycle or playing a musical instrument. This is procedural knowledge that is needed for skills learned by repetitive actions. Procedural knowledge is retained in memory for a very long time and can be retrieved without much effort.

When we assume that knowledge progresses in a continuous cumulation mode, adding repetitions to a habit or a skill enhances its durability in memory and reinforces the "technical competence" of this action or effort. In other words, only by accepting the mode of continuous cumulation can we explain how "practice makes perfect." Each time we perform the activity that requires the specific skill, we are in effect relying on and benefiting from the cumulative impact of *all* the previous repetitions of the activity, sedimented in the memory.[18]

Declarative memory is the assembly place for all the knowledge a person accumulates from childhood to adulthood. By this definition, declarative knowledge would be all the knowledge we accumulate in our lifetime. This is knowledge of facts and, by definition, knowledge which remains in memory for a long term, over many years.[19]

Declarative memory (and knowledge) is classified in two types: *episodic* memory and *semantic* memory. The difference between these types of memory is the degree to which we recall in memory aspects of a personalized experience. Thus, semantic memory refers to general knowledge of facts and names, so that a "meaning" is attached to words and facts, and we know to utilize them to particular circumstances. When we recall facts, names, facial expressions, and other general knowledge, *and* we can relate those to particular times and places (episodes) as a cohesive form of recalling experiences, we are tapping our episodic memory.[20]

A good example is the event of a "sunset." In semantic knowledge it means the slow disappearance of the sun in a fiery ball, in the evening hours, in the west. In episodic terms, "sunset" to me is an event in which I am sitting with my wife on the beach in Mexico, observing the slow descent of the ball of fire into the blue horizon, and the evoking of romantic sentiments.

Semantic knowledge and its deposit in semantic memory allows for the sharing of knowledge among knowers. By providing meaning to facts and pictures, we transfer them and thus overcome "ontological heterogeneity." This means that although the facts and pictures are viewed differently by each knower, there can be a common and shared meaning which allows for transfer of knowledge across the individualized and subjective views of reality.[21] People can therefore communicate what they know and what they remember by assigning meaning to their knowledge in terms that can be shared with others and easily understood by them—even though to them this knowledge may evoke very personalized and unique emotions and memories.

The model of continuous cumulation of knowledge best explains the categories of declarative memory and how knowledge is stored in them and retrieved for recall by the knower. If we examine the evolutionary approach, natural selection of knowledge memorized will mean that it will be as good as the next change or threat from the environment. As new experiences are added, the current cumulative pool will be challenged. The survival of existing knowledge depends on the evolutionary model, on how fast and how well one adapts to the new and changed environment—regardless of how much experience has been collected. If food becomes scarce on the ground, regardless of how well one can maneuver on the ground and the experience accumulated in walking and running, those who are able to climb trees and access the food supply high on trees will be the only survivors.

In accumulating knowledge in our lifespan, we do not opt for the "best" knowledge or that which best fits the current conditions in our environment. This is because of the question: who defines what is best? In biological evolution it is up to nature, but in the case of knowledge, we lack a judge, objective and knowledgeable as well as acceptable, who defines which is the best knowledge to retain and which is the best knowledge to discard. Clearly, in the scientific method we have established procedures that help us decide what is and is not acceptable as *scientific* knowledge. But this method does not extend to all forms of knowledge.

Secondly, to require "best knowledge" would be very inefficient and would tax brain resources. These would have to be engaged in careful and continuous processing of inputs to determine which knowledge is "best"—as criteria would of course continually change. We would also have an added burden to decide which knowledge is *generally* accepted and which would also be transferred to memory.

The Case for Cumulative Knowledge

We define declarative knowledge as composed of semantic and episodic categories of knowledge. I am introducing here the notion of *cumulative knowledge* which is the clustering of semantic and episodic knowledge. In our lifetime we accumulate knowledge (and retain some of it in memory) that is the product of clustering of facts and appearances (faces, photographic descriptions), and the clustering of episodes or events in which these facts and appearances participated in our experience.

For example, in the case of "sunset," I cluster the facts of "the setting of the sun" with "kiss," with "touch," to form the cluster of "romance." The episodic knowledge will be when and where I experienced romance, and the clustering of all episodes in my life in which romance was a defining occurrence.

A different mode of cumulative knowledge is *scientific knowledge*. This is a clustering of facts (semantics) such as: "bacteria, "tissues," "flow," "acids," and "food particles," clustered into a higher-order construct such as "digestive tract." Each person may have a memory of the first experience with the digestive tract (such as its malfunction leading to pain and vomiting), but the scientific knowledge thus accumulated is not linked to episodes in the life or experience of the knower. Scientific knowledge is created and is acquired in an objective mode, as much as possible, so that knowledge is hopefully shared without reference to the unique individual experiences.

As the product of the clustering of semantic and episodic knowledge, cumulative knowledge is the total catalog of knowledge we have acquired in a given timeframe or throughout our life. This has implications for the creation and utilization of databases and knowledge warehouses, which are discussed in Part IV below. But cumulative knowledge also has implications for the relevance of knowledge we accumulate and the phenomenon of obsolescence of such knowledge.

Cumulative Knowledge, Obsolescence, Learning, and Working on the Margin

The continuous cumulation model of knowledge paints a picture in which knowledge progresses in sediments, continually adding layers upon layers. Cumulative knowledge over a time period generates a large base or catalog of what we know. In the allegory of the wall, we are continually erecting the wall.

This phenomenon, as described by the model, leads to three annoying questions: (1) Why do we need to accumulate and remember so much knowledge? (2) Will not some or much knowledge we accumulate become obsolete? (3) And how will all this knowledge we accumulate help us learn and what will be the role of obsolescence in our learning processes?

These are excellent questions and I regret having asked them because they now require three different answers and a reexamination of the model of the progress of knowledge. Well, not necessarily, since perhaps we can answer all three questions with a lens of the analysis of the progress of knowledge as a sedimentary process.

We accumulate vast amounts of knowledge and we do so in a cumulative or sedimentary manner primarily because such a catalog of knowledge is an instrument in our survival. Whether we are sophisticated urbanites in a high-technological society or hunters-gatherers in the early days of civilization, we need a substantial volume of knowledge to perform those activities that are necessary for our daily lives and for our survival. The knowledge base we accumulate allows us to successfully execute the functions of social beings. The more facts we know and the more experiences we accumulate of episodic knowledge, the more we facilitate not only our power to make decisions in a hostile environment but also to learn.

Our accumulated knowledge base can be classified into two categories: *basic* and *transitory*.[22] The *basic* catalog of knowledge is the bulk of cumulative knowledge. Since early childhood we gather knowledge about ourselves and our world. We know how to function in our surroundings and how to behave in our social setting. We gain knowledge about communicating with others (reading and writing), and we accumulate more esoteric knowledge about computing and the principles of matter and of our universe. This is a base of knowledge that is crucial for our continuous survival.

The *transitory* type of knowledge is primarily technical and scientific, geared toward an activity in the economic or social environment in which we function. For example, we know how to operate a manufacturing machine or how to write a computer program in a certain computer language. As the machine and the language become outdated, our knowledge is now obsolete. But the basic knowledge we had accumulated (such as knowing how to read, write, add and subtract, and how to care for the young) is still very much current and necessary for our continuing survival.[23]

We may therefore argue that the phenomenon of obsolescence occurs at the *margin* of our knowledge base. This is also in line with the cumulation model. Recent sediments we keep adding to our knowledge base may become obsolete, but the bulk of the knowledge base will continue to matter and to be of use in our daily lives.

We need the massive cumulative knowledge in order to deal with challenges and with new situations we encounter. Therefore, learning may also be classified as basic and transitory. The former is meant to be "life learning," without an expiration date, whereas the latter is learning knowledge which we accept as being transitory and in need of replacement sometime in the future.[24]

The continuous cumulation model explains how we add layers of knowledge to a massive base. Some of these layers will become obsolete and we need to add other layers. We do not really replace obsolete knowledge. It is still part of our catalog of what we know, but of less utility in our struggle for survival in daily life. Our task is to add knowledge that will be more useful, in lieu (but also in addition) to that which is obsolete (Geisler, 2006).

In the next chapter I expand the model of progress of knowledge to make several inquiries: how knowledge continues expanding and how we can see many phenomena in our lives and in our universe via the prism of knowledge. These are fundamental questions, so that answering them makes it possible to use the progress of knowledge as a lens through which so much can be explained.

References

Geisler, E. (2001). Good-bye Dodo bird (Raphus cucullatus): Why social knowledge is cumulative, expansive, and nonevolutionary. *Journal of Management Inquiry, 10*(1), 5-15.

Geisler, E. (2006). A taxonomy and proposed codification of knowledge and knowledge systems in organizations. *Knowledge and Process Management, 13*(4), 285-296.

Leydesdorff, L. (2001). *A sociological theory of communication: The self organization of the knowledge-based society.* Boca Raton, FL: Universal.

Uren, V., Shum, S., Mancini, C., & Li, G. (2004, 22-27 August). Modeling naturalistic argumentation in research literatures. *Proceedings of the 4th Workshop on Computational Models of Natural Arguments,* Valencia, Spain.

Endnotes

1   See, for example, Lehaney, B. et al. (2004). Beyond knowledge management. Hershey, PA: Idea Group. In this book the authors link knowledge management to socio-technical systems (Chapter III) and to systems thinking (Chapter IV). Other examples include: Leydesdorff (2001); Poovey, M. (1998). A history of the modern fact: Problems of knowledge in the sciences of wealth and society. Chicago: University of Chicago Press; Lang, E. et al. (1991). Modeling spatial knowledge on a linguistic basis: Theory, prototype, integration. New York: Springer-Verlag; Yang, C. (2003). Knowledge and learning in natural language. New York: Oxford University Press; Barber, A. (2003). Epistemology of language. New York: Oxford University Press.

2   In Geisler (2001), I compared the growth of knowledge to the extinction of the Dodo bird and had argued that social phenomena are unique and different from physical phenomena. Social phenomena are temporary and highly transitory. They occur when "a set of circumstances, actors, and interactions is congealed in a point in time and space…never to be precisely duplicated."

3   See a more detailed analysis in Chapters X and XI of this book.

4   See, for example, Calvin, H. (2004). A brief history of the mind: From apes to intellect and beyond. New York: Oxford University Press; Zimmer, C. (2004). Soul made flesh: The discovery of the brain and how it changed the world. New York: The Free Press; Calvin, H. & Bickerton, D. (2000). Lingua ex machina: Reconciling Darwin and Chomsky with the human brain. Bradford, UK: Bradford Books; Geary, D. (2004). The origin of the mind: Evolution of brain, cognition, and general intelligence. Washington, DC: American Psychological Association; Sterelny, K. (2003). Thought in a hostile world: The evolution of human cognition. Oxford, UK: Blackwell.

5   See the discussion of consciousness in Chapter VI of this book.

6   Scientists are increasingly discovering that certain nutrients help in human "thinking" and "memory." These nutrients do not act on memory itself. They are instrumental in making the factory more effective and productive, as an improved energy flow and distribution would improve the performance of a manufacturing plant.

[7]   Greenspan, S., & Greenspan, N. (1985). First feelings: Milestones in the emotional development of your infant and child from birth to age 4. New York: Viking Press; Greenspan, S. (1997). The growth of the mind and the endangered origins of intelligence. Reading, MA: Addison-Wesley Longman.

[8]   There is a possibility of explaining many social and educational phenomena by the way in which knowledge progresses and is used. This example of early childhood development is a good example.

[9]   As I repeatedly explained in previous chapters, this allegory is also a simplification of a complex phenomenon. It is given here as a didactic aid, with the purpose of illuminating the subject matter with a simple allegory. This effort is similar to Plato's allegory of the cave. There is also an issue with the Sorites Paradox, described in the previous chapter.

[10]   In Geisler (2001), I described the progress of social knowledge. The same principle listed in that paper also apply to all human knowledge.

[11]   Synergy is not a phenomenon that helps knowledge to survive a selection process. It is simply a statement that different types of knowledge may complement each other, thus enriching the entire pool. This is especially true with cross-disciplinary benefits.

[12]   Reconfiguration of existing knowledge is in line with models of scientific pathbreaking such as Kuhn's model of revolutions in science.

[13]   The reader may wish to consult: Uren (2004, pp. 34-42) and Sasaki, M. et al. (1998). Quantum channels showing superadditivity in classical capacity. Physical Review, A58, 146-158; Sohma, M., & Hirota, O. (2001). Information capacity formula of quantum optical channels. Quantum Physics, 10(2), 1-20; Giovannetti, V., & Lloyd, S. (2004). Additivity properties of a Gaussian channel. Physical Review, A69, 42-50.

[14]   We assume a relatively unhindered growth potential of the knowledge pool in the human brain.

[15]   These two types of analyses can be done for the individual knower and for collective cumulation (such as databases and knowledge warehouses).

[16]   Done by the use of electroencephalograph (EEG) or magnetoencephalograph (MEG), which also identifies the location of the electromagnetic activity in the brain.

[17]   It should be clarified that the electrical charges of these brain waves is very weak. We generally estimate that a fully functioning human brain, where all ten billion neurons are excited and processing knowledge, would only generate up to 1-5 volts of electricity. As an illustration, the call of the whale is usually in a range of frequency of 15 and 20 Hertz (human beta waves).

[18]   Procedural memory contains knowledge that is "knowing how, whereas declarative knowledge is knowledge that (facts). A good example of the role of previous knowledge is the case of being aware of venereal diseases or AIDS. When one has unprotected sexual relations, one in effect has them not with the current partner, but with all the sexual partners he/she had before. This is how knowledge also progresses and is the "essence of experience."

[19]   There has been substantial literature on declarative memory. The interested reader may wish to consult: Rubin, D., & Schulkind, M. (1997). The distribution of autobiographical memory across the lifespan. Memory of Cognition, 25(4), 859-866; and the original work by Tulving, E. (1972). Organization of memory. New York: Academic Press. Also, Tulving, E., & Schachter, D. (1981). Primary and human memory systems. Science, 247, 301-306; Nolte, J. (1981). The human brain: An introduction to its functional anatomy. New York: Elsevier-Mosby.

[20]   See, for example, Tulving, E. (1984). Precis of elements of episodic memory. Behavioral and Brain Sciences, 7(2), 223-268; Delacour, J. (Ed.). (1994). The memory system of the brain. New York: World Scientific; Eichenbaum, H. (1997). Declarative memory: Insights from cognitive neurobiology. Annual Review of Psychology, 48(3), 547-572.

[21]   The reader may be interested in the topic of information and knowledge integration as means to overcome diverse perspectives in semantics. For example, Firat, A., Madnick, S., & Grosof, B. (2002, October). Knowledge integration to overcome ontological heterogeneity: Challenges from financial information systems. MIT Sloan Working Paper 4382-02.

[22]   I do not wish to encumber the reader with additional classifications. However, the categories of basic and transitory are central to the explanation of how obsolescence occurs on the margin of our knowledge base, rather than on the entire collection.

[23]   In the literature on obsolescence, see for example the classical study: Pakes, A., & Schankerman, M. (1979, May). The rate of obsolescence of knowledge, research gestation lags, and the private rate of return to research resources. NBER Working Paper, W0346.

[24]   The notion that the human knowledge base (and that of other living species) is needed for survival in hostile environments does not at all suggest that knowledge progresses in an evolutionary manner. Rather, knowledge is an instrument in the struggle for survival, but itself progresses in a continuous cumulation mode, adding sediments to a fixed catalog of knowledge collected over long periods of time, even over a lifelong existence.

Chapter X

The Marvels of Analogy:
Expansion and Infinities and Other Matters of Human Knowledge

Expansion and Infinity

Within the limited human existence, the knowledge we accumulate expands indefinitely. For every sediment of knowledge we add to our base, a relatively immense volume of other potential knowledge may be explored and accessed. Neuroscientists agree that we utilize only a small portion of our brains in the processing of knowledge. We are limited by our capacity to explore the environment with our senses and constraints by our brain's ability to process, store, and retrieve the knowledge we do absorb.

Knowledge progresses in continuous cumulation in a way quite similar to the model of the expanding universe.[1] We keep on adding sediments to our knowledge base, with no visible or planned end in sight. There are even several anecdotes about human vanity and lack of understanding of this phenomenon. In the history of technological innovation, scholars are amused by the words of the director of the patent office in New York, who in the early years of the twentieth century declared that all that could be invented had already been invented.

Knowledge, like matter in the universe, has the potential to expand indefinitely. This phenomenon seems to be a guiding principle for matter as well as for living organisms. How is this possible? Consider the accepted notion that biological evolution involves the continuous struggle between species—against their predators and other organisms that exist in their environment and who consume limited resources necessary for the species survival. Natural selection dictates that even within the species, those who better adapt will survive. The generation and use of knowledge by these biological species is a crucial tool in their struggle for survival. As an integral part of their existence and struggle, biological organisms have a very strong drive to perpetuate the species. Living organisms possess the overarching desire to reproduce and to replicate their structure and characteristics with an eye toward eternity.[2]

But, why do these biological organisms so ardently strive to perpetuate themselves?[3] There are four possible explanations. The first is the *genetic* explanation. The desire to procreate and to perpetuate the species is embedded in their genetic make up. Living organisms simply follow the instructions imprinted in their structure.[4] As the organism is formed and grows (following the genetic make up), it is also endowed with a trait or a preprogrammed and instinctive desire to perpetuate its existence.

Why do these organisms (and humans of course are among them) have the desire to perpetuate their existence? Why do they feel the basic need to procreate? Are they selfish, or perhaps vain? Another plausible explanation is that these organisms are somehow *aware of the limits of their existence and of their mortality.* At the atomic level and below, all matter in the universe, biological and inanimate, is the same, but inanimate objects do not procreate—only living organisms (animals and plants) do so. But, do plants and bacteria have an awareness of their mortality? Do they procreate and propagate their species because they wish to overcome the inevitable destiny that awaits them, that is, a limited existence and a predetermined span of life? Clearly, the answer would be that these organisms do not (to our knowledge) have such understanding. We humans have such a cognitive ability, perhaps, because we are in possession of knowledge about life, mortality, seasons in the environment—and the ability to draw conclusions about our own mortality.

A third possibility is that if, as physical science tells us, the amount of matter and energy is fixed so that we can transform one into the other but not add to them, *the universe is expanding* in this very way. As organisms propagate, they consume resources and struggle with each other so that along the way there are some periods of balance and the expansion is moderated.[5]

This is analogous to the problem of what came first: the chicken or the egg? Do biological organisms propagate because they are subjected to the forces of the struggle for existence and for resources they need to survive, or the fact that the universe is expanding and this principle then applies to all living creatures? Is it because the biological ecosystem is such that to maintain a balance there is a need for a continuing flow of biological species, as species use other species for their maintenance and survival? Alternatively, is it because the universe is expanding toward its ultimate extinction, hence all matter follows suit and has a limited life? But, why only living organisms? Rocks and water, for example, do not reproduce and they are subjected to forces of nature, to erosion, and to transformations.

What is so special about living organisms that they possess the urge and the ability to procreate and propagate? A fourth explanation may be because of their complexity (as compared with rock formations) or the fact they can interact with their environment. It seems that the more complex the living organism, the more it develops the abilities to challenge and even to overcome "natural" trends of limited coexistence. For example, humans have created social structures that allow them to provide their participants with care so that life expectancy has been elevated beyond the "natural" limit for humans in their evolutionary context.[6]

So, living organisms are programmed to propagate in order to perpetuate their existence—regardless of whether they are cognizant of their limited existence. The more complex the organism and the more it is able to acquire and to process knowledge about itself and the environment, the more the organism (such as humans) is able to challenge the forces in nature and in its genetic make up that make it vulnerable and mortal.

Therefore, knowledge must be continually generated and processed, so it can serve as an instrument in the struggle against mortality. These complex creatures (we human beings) have at some point begun to comprehend that procreation alone is not a guarantee for perpetuation of oneself or the species. Knowledge is therefore being accumulated and utilized to improve procreation (hence future generations) *and* to improve the existing generation—also as a means for continuation into infinity.[7]

Why We Search for Knowledge

Human beings are finding themselves on a quest for knowledge. We do so because it helps us to cope with an increasingly dynamic world and, as a consequence of millennia, of struggle for survival.[8] Alas, we also search for knowledge because of the thrill and excitement the search effort provides, and because "It's there!"

Because of the continuing expansion of knowledge, we are unable to satisfactorily measure its standards, its benchmarks, and its levels of sufficiency. This means that we continually accumulate knowledge about a very large number of topics and areas of inquiry, but we cannot tell when and where we reach a point of sufficiency or saturation. Knowledge is expandable, for all practical purposes, into infinity.[9]

In a way we are caught in a "Knowledge Race," in which we race against each other, against time, and against obsolescence of some of our knowledge base. In some scientific disciplines and technical fields, we may reach an agreement—by consensus—that we have the quantity and quality of knowledge we need, and that it is time to "abandon the search" and to move on to a different area of inquiry. But, in general, we are on a relentless quest for knowledge that has no frontiers nor limits, extending into an inexhaustible pool of potential knowledge about ourselves and about our external environment.

We are also discovering that our brains are in tune with the limited process of gathering knowledge and sharing it with others through communication and language. In 2004, neuroscientists discovered empirically that the brain has a third area involved in language ability. This area is named after Norman Geschwind (1926-1984) who hypothesized its existence. Neurologists also discovered that this area of the brain matures later in life, at the time when the child develops reading and writing abilities (see Catani, Jones, & Ffychte, 2005). Thus, the need to keep up with the processing and communication of knowledge obliges the brain to allocate additional and targeted space for this task.

The Progress of Knowledge and the Modern Age

In recent years a special chapter has been written on the generation and utilization of knowledge. This chapter is the culmination of over a century of what is usually described as the *modern age*. A small set of critical factors, combined with several trends, have produced the current state of modernity and the knowledge society and economy. But, what is this contemporary state of our economic and social life that we call modernity, and what role does knowledge play in it?

At the outset, we need to better define the terms "modernity" and "postmodernism" while avoiding complex sociological terminology. *Modernity* or modernism may be described as a social and economic development of human existence based on industrialization. This stage of development is also marked by profound scientific and technological innovations which permeate almost any aspect of our lives.

Postmodernism is a term that became popular since the 1980s, and it generally refers to more recent trends in what is sometimes called "the post-industrial" society. In 1976 Daniel Bell listed five dimensions of post-industrialization: (1) movement from producing goods to a service economy, (2) the rise to dominance of a professional and technological class, (3) growing centrality and importance of knowledge, (4) strong orientation toward the future, and (5) intellectual technology or decision making increasingly based on science and knowledge (the "Knowledge Age").

Since it is difficult to accurately establish the point in time in which modernism became postmodernism, the focus here should be on the role of knowledge in the contemporary society and the economy of the twenty-first century.[11] We find a healthy debate and strong disagreement on what exactly is this new "age of knowledge," who are the "knowledge workers," and how significant are the changes in our life because of this knowledge "revolution."

In his various publications and lectures, Peter Drucker argued that we are witnessing the emergence of the knowledge society, fueled by knowledge workers as a dominant force in it. Knowledge works, Drucker (1994) said, "even though only a large minority of the work force, already give the emerging knowledge society its character, its leadership, its central challenges, and its social profile."

Drucker and others also defined knowledge workers as employees, working for organizations. The knowledge society is therefore a "society of organizations" where the work and its performance are dependent upon the collective effort and the collective knowledge of its employees.[11] Contemporary workers are in effect "merchants of knowledge," selling their ware to the highest bidding organization, and swiftly and seamlessly moving from one employer to another. They offer the new resource in addition to, or in replacement of, capital and land. If, in the past, knowledge was needed to support land and capital to form the industrial society, in the post-industrial era, land and capital support knowledge. Examples are hospitals supporting physicians and the venture capitalists that support entrepreneurs with the ideas and the knowledge to create new businesses (see Harvey, 1989).

Knowledge in this context includes technical and scientific knowledge, social and managerial knowledge, and general knowledge such as that which is acquired throughout high school. By the time today's worker graduates from a university, the catalog of knowledge in his or her possession surpasses all that the highly skilled professionals had acquired during their lifespan a century ago. Moreover, today's knowledge worker must continue to accumulate knowledge in order to combat obsolescence.

We accumulate knowledge in the contemporary society partly in order to reduce uncertainty and to gain a better sense of control over our dynamic environment. Traditional theory of management had proposed that management is the effort to reduce uncertainty. This is true for our quest for knowledge, the need to better understand our surroundings, to be able to better compete and even to predict trends and events, and thus be able to act on all of them.

As the only species that tinkers with evolutionary forces, we came to realize that by accumulating knowledge we unlock some of the secrets to how the universe and ourselves are structured, how we function, and why we do not last. In this quest we have constructed an arrangement of institutions and processes that we now call the knowledge society. I believe that this arrangement just seems to be a format that works, so we adopted it and further keep modifying it to accommodate our insatiable appetite for knowledge.

Let us pause for a moment and reflect on how our contemporary society is shaped by knowledge, knowledge workers, and the constant pursuit and accumulation of knowledge. Just as in past generations people expended effort to accumulate land and capital, today we expend resources to accumulate the new source of wealth that fuels our economy and is the essential ingredient of our society: knowledge. This topic will be discussed in Chapter X, in conjunction with implications on databases and knowledge warehouses.

Summary

Knowledge progresses in a continuous cumulation of clustering of sensorial inputs. As we collect, store, and process this ever-growing catalog of what we know, we make many uses of it, so that the knowledge we have gained impacts the way we live. In the next chapter I offer an optimistic view of the progress of knowledge. But, the main issue that I will bring up in the next chapter is the opportunities offered by the mode in which knowledge progresses. As we add layers to our knowledge base and create new configurations of its cumulation, we open the door to countless ways to improve our existence. These may even be good enough to justify the effort we invest in cumulation of knowledge.

References

Bell, D. (1976). *The coming of post-industrial society: A venture in social forecasting.* New York: Basic Books.

Catani, M., Jones, D., & Ffychte, D. (2005). Perisylvian language networks of the human brain. *Annals of Neurology, 57*(1), 86-97.

Dawkins, R. (1977). *The selfish gene.* Oxford, UK: Oxford University Press.

Drucker, P. (1994, May 4). Knowledge work and knowledge society: The social transformations of this century. In *Proceedings of the Edwin L. Godkin Lecture.*

Harvey, D. (1989). *The condition of postmodernity: An inquiry into the origins of cultural change.* Oxford, UK: Blackwell.

Endnotes

[1]   I made this analogy originally in: Geisler, E. (2001). Good bye Dodo bird. *Journal of Management Inquiry,* (March).

[2]   See, for example, the theory advanced by Dawkins (1977) on the "selfish gene."

3   In some insect species (such as the cicada), the organism is dormant for 17 years, and awakes only to procreate, lives a very short existence, and dies. The entire "mission" or goal of the species is to procreate.

4   These instructions may be considered a form of knowledge that is transmitted across generations. This type of knowledge does not require processing by the brain, as it is used by the organism to form itself and is therefore "processed" or executed automatically.

5   Scientists who study "sustainable development" have developed models and theories that describe these trends.

6   In the United States, for instance, in the twentieth century, life expectancy almost doubled due to a mix of factors such as improved hygiene, availability, and improvement in healthcare delivery, better nutrition, preventive healthcare, and other such factors.

7   The former Surgeon General of the United States, Dr. C. Everett Koop, once quipped that the "baby boom" generation has decided that death is not an option, hence this generation demands—and is willing to pay for—all the possible benefits, even miracles, that medical science and medical technology can produce.

8   Jane Mattisson's book on the work of Thomas Hardy is an excellent example of an analysis of Hardy's description of the industrial revolution. Hardy's novels examine the changes in English society when the knowledge needed for survival in the industrial and urban environment was very different from that required in the rural and agricultural society of previous centuries. See Mattisson, J. (2002). *Knowledge & survival in the novels of Thomas Hardy.* Lund, Sweden: Lund University Press.

9   Notwithstanding such phenomena as "diminishing returns" and patterns of decay and inertia as the knowledge base expands. Even if these phenomena (or principles) do apply to the human knowledge base and my model of progress, they do not negate nor diminish the progression of knowledge and its expansion without a planned, acceptable, or measurable end in sight. These principles, if they apply to the progress of knowledge, merely suggest that the *utility* of some additional knowledge may not be as high as other items of knowledge in the existing knowledge base. See, for example, Munz, P. (1985). *Our knowledge of the growth of knowledge: Popper or Wittgenstein?* London: Routledge & Kegan.

And in the specific case of management, see Sahlin-Andersson, K., & Engwall, L. (2003). *The expansion of management knowledge: Carriers, flows, and sources.* Stanford, CA: Stanford University Press.

[10]   There is substantial literature in various disciplines (such as sociology, anthropology, psychology, and economics) on the role of knowledge in contemporary life. See, for example, Leet, M. (2004). *After effects of knowledge in modernity.* Albany: State of New York University Press; Toulmin, S. (1990). *Cosmopolis: The hidden agenda of modernity.* Chicago: University of Chicago Press. Also see Zuboff, S. (1988). *In the age of the smart machine.* New York: Basic Books; Diggins, J. (1995). *The promise of pragmatism: Modernism and the crisis of knowledge and authority.* Chicago: University of Chicago Press.

[11]   The definition of knowledge workers is at best fuzzy, because it is an indirect approach to defining and describing the "knowledge society," which in itself is also fuzzy. There is not a precise definition of the contemporary "knowledge" that has generated such transformations in society and the economy. Rather, as producers' analysis illustrates, the description focuses on the transformation of workers from craftsmen to hired help in the factory, to their contemporary status as contributors of the resource of knowledge to their organization.

Section III

Epistemetrics:
The Metrics of Knowledge

Chapter XI

Epistemetrics:
The Metrics of Knowledge

Epistemetrics: What We Measure

This chapter introduces the nation of "epistemetrics," the metrics of knowledge. This is a conceptual space which consists of the measurement of the attributes of knowledge, including its origins, processes, flow, and the assessment of its value to users. Epistemetrics is a subfield of the study of knowledge and knowledge systems. It is coined here as the encompassing definition of the measurement of the phenomenon of knowledge.

Epistemetrics contains three complementary topics or sections: (1) *what* we measure, (2) *how* we measure, and (3) *why* we measure. The frame of reference of the concept and its three elements listed above distinguish it from the conception of epistemetrics described by Nicholas Rescher (2005). His conception of epistemetrics is derived from his prolific work in the philosophy of knowledge (Rescher, 2003). The key to Rescher's definition and scope of epistemetrics is his focus on the role for the methods used to link scientific conjectures with experiential data. Rescher's definition of epistemetrics is not about measurement and metrics—unlike the notion of epistemetrics which I employ in this book as the "*metrics* of knowledge."

Rescher's book explores the limits of knowledge. He discusses scientometrics as the main component of his version of epistemetrics. The use of Rescher's approach is philosophy of science and epistemology.[1] This is fundamentally different from the notion of epistemetrics I am proposing in this book. The *metrics* of knowledge is a notion particularly focused on measuring what is human knowledge, how it progresses, and what value is derived from its generation and usage. This conception of epistemetrics is, of course, much broader than any one notion hitherto proposed by Rescher or others (e.g., Kuk, 2006; Ackerman, 2004; Rescher, 1989; McElroy, 2002).

Units and Levels of Measurement

The measurement (metrics) of human knowledge may occur at different levels of abstraction and conceptualization. This is similar to the biological classification scheme of the levels of the subcellular, cellular, and molecular level of biological tissue.[2]

The first measurable unit of knowledge is the clustering of sensorial inputs. This basic unit is the perceptible distortion of sensorial inputs where intervals are perceptible and significant so that they form a measurable entity which is then "known" to the knower and constitutes the fundamental unit of knowledge.

KANEs (*K*nowledge b*A*sic u*N*its of *E*xistence) are the next level of abstraction and measurement of knowledge. They are similar to the cellular form of life, namely, the smallest unit of living matter which is able to function independently—the basic unit of life.

As in biology, the mind congregates units of knowledge to form higher-order constructs which create unique representations of the environment of the knower—hence generating knowledge about the world of the knower—what it looks like and how it operates. Different configurations, clustering, or combinations of the basic units of knowledge create different perspectives of the same phenomenon in the environment. The same knower may have different perspectives of the same reality over time, and different individuals may have simultaneously different knowledge of the same environment.

Clustering units of knowledge into higher-order constructs forms a "hierarchy of complexity." As the mind constructs these ever more complex constructs, it creates a richer and perhaps more viable representation of nature or reality. It also facilitates the transcription, exchange, and transfer of knowledge among individuals.[3]

KANEs are further clustered to form constructs of increased complexity. As cells form molecules, so do KANEs cluster to create "intellectual nuggets." Molecules are defined as the smallest form of a substance that has the characteristics or properties of this substance. "Intellectual nuggets" are such *molecules of knowledge.*

Although KANEs are still a challenge to our attempts to measure them, *nuggets* appear in a form that allow for their transcription and measurement with selective ease. Rubenstein and Geisler (2003) defined nuggets as compound statements, serving as units that carry the knowledge embedded in them. They resemble such means of transport as automobiles, train cars, airplanes, ships, and barges that carry transcribed knowledge in the highways, byways, and skyways of our means of communication, such as language. Common examples are statements in the form "IF x ... THEN y."

If intellectual nuggets are the mode of transport by which knowledge is communicated, what is the knowledge content in them? This content is the KANEs, clustered within the nuggets. Thus, the variables x and y in the compound statement "if x ... then y" are the content knowledge composed of a cluster of KANEs. In the example of biology, nuggets are equivalent to tissues that retain qualities of its components yet have a more complex structure and serve a specific function such that the components cannot serve by themselves. Similarly, KANEs compose a nugget but by themselves cannot serve the function of conveying the knowledge they contain.

Intellectual nuggets, for all practical purposes, are the first "truly measurable" forms of knowledge. KANES—at this stage of our human and technological capabilities—are not measurable as units of knowledge. We may have located their appearance in selected regions of the brain, but we still lack an understanding of how they are formed and how they cluster to create intellectual nuggets.[4]

Rubenstein and Geisler (2003) also coined the term "*supernuggets*" to denote items that are tied specifically to the expressed or implied needs and issues of an organization. This is a higher level of measurement which is more aptly covered in Chapters XIV and XV of this book.

In this context, the notion of a *meme* is better related to that of a supernugget. Three decades ago Richard Dawkins (1975) introduced the term "meme" to describe items of information (or knowledge) that replicate themselves among humans, in a way similar to that of the replication of biological genes. This facet of "cultural evolution" has since then attracted the attention of several scholars. More recently Aunger (2002) introduced the term "neuromemetics." He suggested that memes are derived from self-replicatory electric exchanges within the human brain.

Aunger descends from the level of memes as more complex social and cultural constructs to the more elemental activity in the human brain. Aunger and other memetic scholars have not explained the relation between memes and the origins of knowledge. Memes are generally viewed as intellectual or social concepts which spread like a virus in an ever-growing pace among humans who are exposed to it. These memes are ideas and social, cultural, and ethical concepts. They represent higher-order constructs that are generally related to *social,* hence *organizational* needs and processes. Therefore, they may be closely related to supernuggets.

In order to adequately understand how knowledge is created and communicated, we must delve—as I have done in this book—into the level of the elemental components of that which we call knowledge. From there we advance up the conceptual hierarchy to higher-order and more complex concepts. But knowledge is initiated at the elemental level, so it must be measured, at first, at the level of the clustering of sensorial inputs.

What We Measure: A Taxonomy of Knowledge

In addition to classification by level of abstraction, we can also classify knowledge in several schemes. Some taxonomies, for instance, distinguish between knowledge of skills (such as tying shoe laces or mastering a computer: *how* to do something), knowledge about the universe (*where* am I and *what* does the world look like), and knowledge about higher-order social and cultural concepts (*who* am I, and what are the principles by which we live, such as deity, ethics, and morals).

In Geisler (2006) I expanded the classification of knowledge nuggets to include the criteria of *structure, purpose,* and *function* (SPF). The criterion of structure refers to how knowledge is framed and what it contains. Examples of these attributes are generality of the intellectual content of an item of knowledge, uniqueness of the content, and whether the item is a description of a fact or an opinion (conjecture). This scheme yields eight sub-groups, such as: (1) simple-general-factual (SGF), and (2) complex-unique-subjective (CUS).

The first sub-group (SGF) is a knowledge nugget expressed in a *simple* statement and a generalized description of an actual fact. An example would be: "IF the organization fails to hire new people, THEN over time its workforce will be older." This example may be described as "*relative knowledge*" because it reflects commonly accepted items of knowledge that the variables in the statement are correlated.[5]

The second example of a sub-group could be: "IF we instill and propagate among the people of a less-developed country the lofty ideals of freedom and justice, THEN there is a probability of 64% that the people will reject the current repressive political regime." This item of knowledge is complex, unique, and subjective (CUS) and may be called "focused hypothesis."

The eight sub-groups are classes of items of knowledge by format (simple or complex) and content (general or unique/specific; factual or subjective). These types of knowledge are a broad and preliminary description of items of knowledge. What we measure with them is therefore a set of different nuances in the structure and the content of the statements which carry in them the items of knowledge.

The criterion of *purpose* is simply a classification of items of knowledge by "what is this knowledge for?" Another way of classification is to ask: "To what end or purpose have we gone to the trouble of engendering this item of knowledge?" There are three such purposes. We create knowledge for *utilitarian* objectives such as survival, competitiveness, and performance. These purposes are used by individuals and organizations in creating knowledge for utilitarian objectives. Individuals may generate knowledge about skills ("IF I dig a hole in this type of ground, THEN I will find water for me and my family's survival" or "IF I spend 60 hours working with the new computer system in the organization, THEN there is a good chance that I will be promoted").

A second category of purpose is knowledge created for *hedonistic* objectives. Individuals search for and create knowledge to satisfy their desire to know and to experience the pleasurable sensations that acquiring knowledge bestows on the inquisitive human spirit.

The third category under purpose is the creation of knowledge for the objectives and needs of *systems* such as work organizations. Individuals would create knowledge to facilitate their activities and performance within organizations.

Function is the third criterion of knowledge classification. This is the generation of knowledge to obtain or gain certain benefits from the applications of this knowledge to the individual and to the organization. Under each of the purposes described above, there will be benefits or impacts from the knowledge thus generated. This level of classification is described in Chapter XII ("*How* We Measure").

In summary, we measure knowledge from its inception as the clustering of sensorial inputs to the complex form of intellectual nuggets. Biology is the model: measuring life from the subcellular entities all the way to the complex organism. As we advance up the complexity ladders (in both biology and knowledge), it becomes easier to classify and taxonomize the entities under measurement. In the case of human knowledge, the initial forms of knowledge as defined in this book can be described, but are not yet amenable to measurement.

What we measure in the realm of human knowledge is the very personal manifestation of the way human beings gain awareness of their world and their existence. Ultimately we create such knowledge for certain reasons and purposes and to gain certain benefits. But these categories are a more advanced and complex classification of knowledge within social and cultural frameworks. Originally, human knowledge is a very personal experience of the individual's initial interaction with his/her environment—both internal to the body and external. As soon as there are sensorial inputs and the human brain can cluster them, there is knowledge. However rudimentary, in the form of KANEs, this knowledge should be measured.

The Dichotomy of Measurable Knowledge

Measurable knowledge occurs at two levels, one is complex statements and notions. The second is at the microscopic level of the clustering of sensorial inputs in the human mind. These notions are, by their structure and content, more amenable to be shared among knowers. This is equivalent to "explicit knowledge," diffused and shared among humans.

Conversely, knowledge at the original level of the clustering of sensorial inputs poses formidable obstacles to sharing, transfer, and diffusion. This is equivalent to "tacit knowledge." Its main characteristic is the highly personalized attribute of this knowledge as the individualized experience of the knower which is almost impossible to share with others.

I argue here that these two levels or types of knowledge represent distinct phenomena. The tacit form of knowledge is very different from explicit knowledge. The consequence from this argument is the falsity of the commonly held assertion that "explicit" knowledge is the "tacit" knowledge which can be shared and diffused (e.g., Nonaka & Takeuchi, 1995; Nonaka & Nishiguchi, 2001; Nonaka & Teece, 2001). Explicit knowledge is not necessarily the externalization of the tacit knowledge, which is the personalized, experience-based knowledge derived from the clustering of sensorial inputs. We do not go to the same well of tacit knowledge to scoop up some knowledge which can then be shared with others and diffused.

The assertion that "tacit" and "explicit" are different levels or manifestations of the *same* knowledge in a different location is unacceptable. Explicit knowledge is not the public representation of personalized knowledge of the knower. Rather, explicit knowledge is a social or cultural artifact which allows humans to interact and to share, to a very limited extent, what they know—without, however, "revealing" the innermost knowledge they possess as the personalized experience based on the clustering of sensorial inputs.

The argument that there are two distinct types of knowledge helps to explain the issues involved with the various theories and models of knowledge and the structure of the mind. The distinction, for example, between neural network and the ability to manipulate symbols and to generalize concepts may be explained by the existence of two different phenomena. The mind generates knowledge for the knower by clustering sensorial inputs. This, in turn, may be clustered further into more complex, abstract notions which can then be transcribed into symbolic formats such as language, thus amenable to exchange and diffusion by the knower. It is impossible to determine what the portion of "explicit" knowledge is in relation to the "tacit" pool of knowledge—nor is it of any relevance. Since the two types of knowledge are different, they obey different rules and have different attributes and capabilities.

Although "explicit" knowledge and higher-order concepts are originally formed by clustering sensorial inputs into KANEs and continuing cumulation processes, once the higher-order construct is formed, it is no longer a uniquely personalized experience of the knower. It can now be exchanged and diffused and it describes not the individual's experience, but a detached, independent notion of higher-order complexity.

Biology again may be a good illustration of this dichotomy. Microscopic entities such as cells are quite limited in their ability to interact with other cells, whereas higher-order, more complex organisms have a higher facility to interact and to exchange biological matter, as well as knowledge. Such a dichotomy does not negate the fact that complex organisms have their origin in microscopic entities at the cellular and subcellular levels, nor that in order to better understand these complex organisms (such as the human being) it is necessary to examine and to measure their origins at the microscopic level of analysis.[7]

Conclusion and an Illustration

Consider, for example, the concept of "freedom," which is a higher-order construct of symbolic value. This would be what some cognitive scholars term "free generalizations of abstract relations." Marcus (2003), for example, had proposed that the mind is able to simultaneously act as a network of neurons *and* as a manipulator of symbols, so as to allow the mind to engage in the creation of a higher-order abstract concept such as "freedom."

But it is possible for the mind to conceptualize a higher-order construct without having an inherent, albeit separate structure or mechanism which processes and manipulates symbols.[6] The abstract notion of freedom can only be created in the mind as a conceptualization of individual experience. It is created by the clustering of sensorial inputs as any other form of human knowledge. Evoking sentiments or thoughts (as the notion of "freedom" would do) is restricted to the ability of the mind to posit this notion within an individualized framework of what freedom represents, based on the individual's knowledge—which is the result of, originally, the clustering of sensorial inputs.

So what we measure is not the abstract notion of freedom, if we wish to measure *human knowledge,* we would be measuring the individual's conception of this notion—starting with the clustering of the *individual's* sensorial inputs, not the shared or "commonly understood or known" concept.

To the reader this means a dichotomy in the measurement of human knowledge. The knower is faced with the need to measure knowledge at the origins—that is, at the level of sensorial inputs, while also measuring the more complex forms of knowledge, such as intellectual nuggets. Compare this state of affairs with medical diagnosis. At the level of the patient, the diagnosis rests on characteristics of the complex organism (vital signs such as temperature and blood pressure). A more detailed diagnosis explores the microscopic causes of disease (bacteria and virus) as well as the processes at the cellular and subcellular levels.

In the example of "freedom," this notion may be "known" broadly by the knower—based on descriptions by others. However, another level of knowledge would be the *personal* knowledge of what it means to be free, based on the very individualized and non-transferable experience of the knower.

The dichotomy of measurable knowledge described above also suggests another radical conclusion: that there is a need to re-examine our approach to the incorporation of human knowledge into organizational and social systems. The main effort to date has been to improve the transformation of "tacit" knowledge into "explicit" knowledge (e.g., Davenport & Prusak, 1997; Leonard, 1998). The conclusion from this dichotomy is the redirection of effort, from trying to improve transformation of "tacit" knowledge to a newly intensified effort to improve the transfer, exchange, sharing, and transcription of what knowledge can be in fact externalized. The focus should now be on better communicating the body of knowledge that humans can exchange. This will be a body of knowledge composed of higher-order concepts, symbolic constructs, and abstract notions.[10]

References

Ackerman, D. (2004). *An alchemy of mind: The marvel and mystery of the brain.* New York: Scribner.

Aunger, R. (2002). *The electric meme: A new theory of how we think.* New York: The Free Press.

Blackmore, S. (2000). *The meme machine.* New York: Oxford University Press.

Brodie, R. (2004). *Virus of the mind: The new science of the meme.* Allende, MI: Integral Press.

Davenport, T., & Prusak, L. (1997). *Working knowledge: How organizations manage what they know.* Boston: Harvard Business School Press.

Dawkins, R. (1976). *The selfish gene.* Oxford, UK: Oxford University Press.

Elman, J., Bates, E., & Johnson, M. (1997). *Rethinking innateness: A connectionist perspective on development.* Boston: MIT Press.

Geisler, E. (2006). A taxonomy and proposed codification of knowledge and knowledge systems in organizations. *Knowledge and Process Management, 13*(4), 285-296.

Kuk, G. (2006). Strategic interaction and knowledge sharing in the KDE developer mailing list. *Management Science, 52*(7), 1031-1042.

Leonard, D. (1998). *Wellsprings of knowledge: Building and sustaining the sources of innovation.* Boston: Harvard Business School Press.

Marcus, G. (2003). *The algebraic mind: Integrating connectionism and cognitive science.* Boston: MIT Press.

McElroy, M. (2002). *The new knowledge management: Complexity, learning, and sustainability.* Burlington, MA: Butterworth-Heinemann.

Nonaka, I., & Nishiguchi, T. (2001). *Knowledge emergence: Social, technical, and evolutionary dimensions of knowledge creation.* New York: Oxford University Press.

Nonaka, I., & Takeuchi, H. (1995). *The knowledge-creating company: How Japanese companies create the dynamics of innovation.* Oxford, UK: Oxford University Press.

Nonaka, I., & Teece, D.J. (2001). *Managing industrial knowledge: Creation, transfer and utilization.* Thousand Oaks, CA: Sage.

Rescher, N. (1989). *Cognitive economy: The economic dimensions of the theory of knowledge.* Pittsburgh, PA: University of Pittsburgh Press.

Rescher, N. (2003). *Epistemology: On the scope and limits of knowledge.* Albany: State University of New York Press.

Rescher, N. (2005). *Epistemetrics.* New York: Cambridge University Press.

Rubenstein, A.H., & Geisler, E. (2003). *Installing and managing workable knowledge management systems.* Westport, CT: Praeger.

Sorensen, O., Rivkin, J., & Fleming, J. (2006). Complexity, networks, and knowledge flow. *Research Policy, 35*(7), 994-1017.

Endnotes

1   Rescher's book discusses the following conceptual frameworks: (1) Duhem's law of cognitive complementarity, (2) Kant's conception of the nature of knowledge, (3) Spencer's law of cognitive development, (4) Gibbon's law of logarithmic returns, and (5) Kant's notion of cognitive finitude.

2   This classification is different from the data to knowledge continuum. All the levels of measurement discussed here are levels of knowledge—not data, information, or wisdom.

3   This topic is further discussed in Chapter XII: "How We Measure."

4   There is a difference between intellectual nuggets and the notion of "meme." Some memes may contain intellectual nuggets, but not all intellectual nuggets are memes—as these are defined by those who support this notion. For additional readings about memes, see Blackmore and Dawkins (2000), Aunger (2002), and Brodic (2004).

5   Another popular example of SGF-type item of knowledge is the statement from the scientific discipline of chemistry: "IF we bring together, under specified conditions, one atom of the element oxygen and two atoms of the element hydrogen, THEN the resulting reaction will produce a molecule of the substance water."

6   Also see Elman, Bates, and Johnson (1997) and Sorensen, Rivkin, and Fleming (2006).

7   This dichotomy of measurable knowledge is a key concept in the model of knowledge I advanced in this book. The existence of two distinct phenomena—at least from the viewpoint of the scientific examination of what we consider to be human knowledge—engenders several radical consequences and conclusions regarding our perspective on what, how, and why we measure knowledge, and even more so on how we employ human knowledge in complex organizational systems.

10 For further discussion of this topic, see Part IV of this book.

Chapter VII

Epistemetrics:
How We Measure

This chapter is focused on the ways and processes by which we measure human knowledge at both the individual and organizational levels. "How" we measure knowledge is strongly related to the notion of "what" we measure, described in the previous chapter. The nature of knowledge that can be measured is the externalized or explicit knowledge shared and diffused among individuals and their organizations. We recognize the existence of KANEs as the clustering of sensorial input, but we are unable at this point to adequately measure them. We have the capability to measure sensorial activities and the locations in the brain of excitations and activities that signify cognition and emotions, but we are still unable to measure knowledge at the fundamental level of clustering of sensorial inputs.[1]

What we can measure is knowledge in the form of notions and intellectual constructs that can be shared and externalized. There is a relationship between our inability, at present, to measure the fundamental elements of knowledge and our current inability to transfer, share, and exchange such knowledge. At the initial level of clustering of sensorial inputs, there are as yet insurmountable difficulties in transcribing such elements to a medium such as language which allows for communication and exchange.

Paradoxically, the more complex the higher-order constructs of knowledge, the more they allow for transcription and sharing among individuals and their organizations. These transcribed items of ("explicit") knowledge contain only a small portion of the entire stock of knowledge in the human mind. So the limited volume of transferable knowledge and the constraints of communicating and exchanging combine to make it very difficult for individuals to share and diffuse what they know. If we could exchange knowledge at the initial clustering stage—as neurons do at the chemical level—the sharing of knowledge would have been in the form of KANEs and would have resulted in outstanding exchange and almost error-free communication processes.

So, the pertinent issue is not how to improve the transformation of "tacit" to "explicit" knowledge, but how to improve the exchange of knowledge we *can* transfer and communicate. This means we should focus on mechanisms of exchange and transfer: how to measure knowledge and share it.

Key Dimensions

Since the early days of hunters-gatherers, humans have relentlessly practiced their skills of communication. They continually improved the exchange of knowledge for the purpose of survival and the training and mentoring of their children. From these primitive hunters-gatherers to Wittgenstein and Chomsky, the role of communication and knowledge exchange among humans became a key activity and a crucial skill in human experience as a social being. With the focus on the knowledge which can be shared and exchanged, the measurement of how this exchange occurs starts with the transcribed form of intellectual nuggets. The "flow" refers to the conduits or vehicles in which knowledge resides and which serve as means for sharing and diffusion.

We measure knowledge by the modes in which intellectual nuggets are transformed, once they are transcribed and shared. The same modes or conduits for sharing apply to the hunter-gatherer who relates the day's adventures and the prey captured as to the corporate executive debriefing her fellow managers on a complex business transaction (e.g., Irani, Slonowsky, & Shajahan, 2006; Biederman, 1987).

This phenomenon of sharing knowledge consists of three dimensions: *Transfer*, *Repository*, and *Processing* (TRP).[2] The *transfer* dimension includes the vehicles or means of transfer and implementation. Table 1 lists these vehicles. For example, "debriefings," "audits," and "periodic" reports are means of knowledge transfer, whereas "rules," "policy directions," and "cultural tenets" are means of implementing such knowledge in organizational frameworks.

Imagine that we are measuring the types and flow of vehicles carrying the cargo of knowledge. This is comparable to people in rush hour, taking commuter trains, buses, and automobiles to their places of work. We measure the movement of commuters to the point when they enter their workplace. In this context we do not measure the nature of the skills they bring to their job, nor how effectively they put these skills to work.

But, we do measure the implementation of knowledge being transferred. The means of implementation within organizational structures are the mechanisms that people and organizations use to share and diffuse knowledge in a form in which it becomes manageable and can be shared with and absorbed by others (see Gertler & Vinodrai, 2005).

Table 1. Vehicles or means by which knowledge is transferred and implemented

Vehicles or Means of Transfer	Vehicles or Means of Implementation
• debriefing	➢ rules, procedures, and regulations
• audits	➢ policy and strategic directions
• periodic reports	➢ codes for individual and organizational conduct
• mentoring	
• evaluations and assessments	➢ code of ethics and ethical behavior
• statements by experts	➢ cultural tenets
• policy statements	➢ principles of organization
• lectures	➢ principles of management
• lessons	➢ science, technology, research and development methods, directions, and practices
• speeches	
• announcements	➢ evaluation and assessment of people and organizations
• documentation: written and electronic	

The *repository* dimension is the transaction of knowledge in which intellectual nuggets are stored in a form that allows for their repository and retrieval. This dimension has two components. The first is the *structure* of the repository, which refers to the physical form of storage and manipulation, and the architecture for such storage.

The second component is the *rationale* for the repository. This includes the principles, logic, and procedures by which knowledge is stored and retrieved.

Individuals retain knowledge in many forms, as do their organizations. The concern in the repository dimension is the ways and modes of storage. Without the means to keep knowledge overtime and to keep it from deteriorating, being forgotten, or unduly changed and mishandled, individuals and organizations would be unable to transfer and to process knowledge beyond the immediate exchange between individuals. Also, repositories of knowledge allow for a wider distribution and diffusion to a larger number of individuals and organizations. As technologies for storage and transformation emerge and are further perfected (for example, from writing to digital libraries), the advantages of knowledge repositories become more pronounced and ubiquitous (Cooke, 2002).

Knowledge bases or repositories for knowledge are the artificial mechanisms designed to store knowledge outside the human brain, its memory, and its cognition. Clearly, only knowledge that appears in an "explicit" or transcribed mode can be stored and further transacted.

But, a key strength of the mechanisms of storage is not in the volume of knowledge nor the speed or other attributes of the flow of knowledge. Rather, it is the capabilities or facility with which intellectual nuggets in the repository can be clustered for storage and for further extraction by users of the repository. The intellectual nuggets have to be codified and categorized in a manner that will allow for clustering.

Consider knowledge nuggets describing the physical dimensions of each of the planets orbiting the sun. The strength of the repository would be the ability to cluster these nuggets into more complex nuggets describing a "solar system," then creating nuggets which describe attributes of such systems and allow for cumulation with other supernuggets on similar systems to form higher-order constructs of the structure and evolution of stellar matter.

The Dimension of Processes

The third dimension of "how" we measure knowledge includes the processes by which knowledge flows in organizations. These are the processes that move knowledge from the stage of its generation to its eventual utilization by users.

There are two main stages of transfer and transformation of knowledge. The first is from the generators to implementation and absorption. The second is the move to utilization and reuse. Within each stage of the process, there is a repository in which knowledge can be stored, until it is captured again and processed. This model of the processing of knowledge applies to the transfer from individual to individual, and from individual to an artificial mode of knowledge containment and processing, such as computers, knowledge bases, and knowledge systems.

In the case of knowledge transferred to an individual, implementation and absorption means that the knowledge thus captured by the individual is integrated with other knowledge the individual possesses, so that clustering occurs and there is progress towards sense making in the individual's mind and understanding of her surroundings.

Utilization of knowledge by the individual occurs when the knowledge absorbed and implemented contributes to the functions, activities, and goals

Figure 1. General model of processes of knowledge in organizations

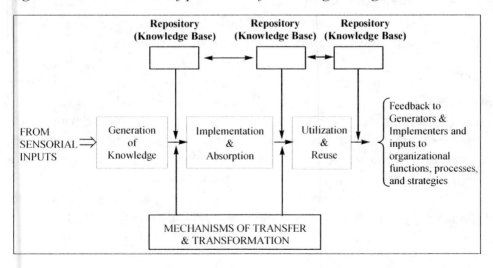

of the individual. A simple example would be the capturing of knowledge about business opportunities in an emerging market. When clustered with the existing body of knowledge about the individual's own capabilities, those of her company, and how to compete in the international arena, such knowledge contributes to the individual's decision to engage the organization in a new venture in the emerging market.

In the case of processes in which there is transfer of knowledge between individuals and knowledge bases or other organizational artifacts, the organization as well as the individual benefit from the process. Consider, for example, the knowledge gained by an individual researcher of the variation in a human gene that may account for about 17% of dyslexia, the reading disability that afflicts a great number of children and adults. This knowledge is transferred to a knowledge base on human genetics and on the disability of dyslexia. Within each repository this knowledge is clustered with existing knowledge to be implemented into activities and goals of better diagnosis and therapeutics of cognitive disorders in general, and dyslexia in particular. As such genetic knowledge contributes to improved diagnostics and therapeutics, this knowledge is utilized and can further be revised in other genetic explorations.

The processes of transfer of knowledge provide us with one aspect of the measurement of knowledge. The transfer and mobility of knowledge from its generation to its utilization is a measure of how knowledge moves between individuals and their organizational mechanisms and artifacts. Cognizant of the fact that not all knowledge is easily transferable, nor effortlessly implementable, we may classify types of knowledge by the attributes of the process by which they are transferred.

Figure 2 shows four possible types of knowledge resulting from the intersect between the level of transferability and the ease of implementation of knowledge. The first type is *improbable,* when the rate of transfer is slow and untimely, and the implementation is difficult. This is a case when knowledge exists, but is not provided to the implementers on time, and in a form which makes it harder to incorporate it into the existing knowledge base. This is, for example, the case of the pre-September 11, 2001, intelligence on suspected terrorists operating in the United States. Although the knowledge existed within the intelligence community, its transfer and sharing was done in an untimely manner, and in formats which made it impossible to incorporate into existing knowledge bases on international and domestic terrorists.

Figure 2. The intersection of the attributes of knowledge of transfer and implementation

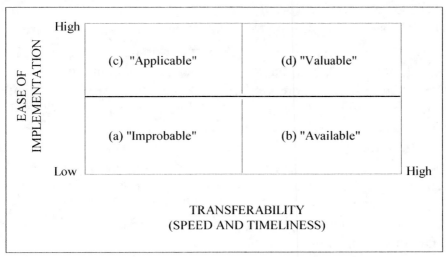

The second type is *available,* where knowledge is transferred in a timely fashion but lacks the attributes that allow for easy implementation. This is knowledge available but not easily clustered or implementable. For example, during the Second World War, the British General Sir Neil Ritchie was facing the German "Africa Corp" under the command of Field Marshal Erwin Rommel. In late May 1942 Ritchie received substantial knowledge about Rommel's battle plan, as deciphered by British Intelligence and transferred to him in a timely manner. For several reasons, among them Ritchie's distrust of the source of the knowledge, he failed to implement this knowledge and fell into Rommel's trap, lost the battle and 300 of his tanks. The speedy and timely transfer of knowledge makes it available but not necessarily implementable.

A third type of knowledge is *applicable.* This is knowledge poorly transferred in a slow and untimely fashion, but in a format which renders it highly implementable. For example, knowledge about the in-situ situation in the aftermath of hurricane "Katrina" in the Gulf states had been transferred to decision makers in a slow and untimely manner, although such knowledge about the gravity of the situation had been highly implementable.

Finally, the fourth type is *valuable,* describing knowledge which is transferred speedily and in a timely fashion and is also highly implementable. The four types of knowledge apply to the exchange of knowledge from: (1) individual to individual, (2) individual to organizational artifact (e.g., knowledge system or knowledge base), and (3) artifact to artifact (e.g., knowledge system to knowledge system–organization to organization).

We measure the dimension of processes as the way in which knowledge is exchanged, transferred, and shared among actors. This third component of the TRP of "how we measure knowledge" offers an understanding of the movement of knowledge among actors. In summary, the TRP (*Transfer, Repository,* and *Processes*) construct focuses on the *how,* rather than the content of the knowledge being exchanged, or the value and utility of such knowledge. Some critics of the emerging field of knowledge management have emphasized this aspect of the phenomenon of knowledge in human and organizational interactions (e.g., Wilson, 2002). In a biased perspective they view knowledge primarily, if not solely, as processes, and the role of knowledge in human interaction as mechanisms of exchange, without paying the necessary attention to the content ("what we measure") and the generation of knowledge—nor to its value and utility ("why we measure")—as three complementary approaches to a truly comprehensive view of knowledge and knowledge management.

Similarly, scholars in the emerging field of knowledge management also tend to emphasize the role that processes play in our understanding of knowledge and knowledge management (e.g., Davenport & Prusak, 1998; Tsoukas & Vladimirou, 2001). They do so partly because of the need for relevancy of the research in this infant area of exploration, and partly because we still lack theoretical models that allow for a more comprehensive view of knowledge—from generation to utilization. The concept of *epistemetrics* is a comprehensive approach containing the "what," "how," and "why" we measure knowledge.

Strengths, Limitations, and Applications

The model of "how we measure" knowledge consists of the TRP components. The key strengths of this model are the comprehensive coverage of the complex phenomenon of the transfer and sharing of knowledge and the integrative framework of the vehicles, repositories, and processes under the unifying notion of TRP. The first strength of the TRP model is the combined approach to the transfer of knowledge. The model includes all the measurable components: the vehicles, the repositories, and the processes. This is a unique perspective where all three components that describe and measure the transfer of knowledge are described within a simple model.

The second strength is the integrative approach, whereby the three components are presented within the overarching model of "how we measure" knowledge. They are also conceptually and methodologically linked to the first element of epistemetrics: "what we measure."

There are two limitations to this model. The first is the potential difficulty in isolating each transfer activity of knowledge—for example, the process by which a given intellectual nugget is transferred (via which vehicle or mechanism), combined with the repository when it is deposited, and the process of implementation. This may generate some design difficulties in the course of empirical research of individual nuggets.

The second limitation involves the integration of the processes in the model with existing knowledge systems in organizations. How do we seamlessly measure the transfer and implementation of "new" knowledge to existing systems? A complementary question would be: How do we absorb such new knowledge within existing practices of the organization? Further research is needed to explore these limitations.

There are three possible areas in which the TRP model of "how we measure knowledge" may have potential applications. The model may be used to categorize knowledge transacted among individuals and between individuals and organizations. In this case, "valuable" knowledge will be given priority in transferring between actors. Conversely, "improbable" knowledge will be destined for transformation and repair so that it may become more "applicable."

Another potential application would be a better distinction made between the stages by which knowledge is transacted. There are differences in goals, procedures, and activities between transfer and deposit of knowledge in appropriate repositories. These distinctions would contribute, for example, to the application of different criteria for the evaluation of the stages.

The third possible application is the model, providing a conceptual framework for the empirical study of knowledge transacted in organizations. By providing distinct stages and measures for these stages, the model allows for a more rigorous empirical investigation of how knowledge flows in organizations. A good illustration would be the need to explore the transaction of knowledge in conditions of uncertainty following a national disaster. The TRP model allows for a coherent dissection of the movement of knowledge through the stages of transfer and implementation.

References

Biederman, I. (1987). Recognition-by-components: A theory of human image understanding. *Psychological Review, 94*(2), 115-147.

Cooke, P. (2002). *Knowledge economies.* London: Routledge.

Davenport, T., & Prusak, L. (1998). *Working knowledge.* Boston: Harvard Business School Press.

Dutta, S. (1997). Strategies for implementing knowledge-based systems. *IEEE Transactions in Engineering Management, 44*(1), 79-90.

Gertler, M., & Vinodrai, T. (2005). Learning from America? Knowledge flows and industrial practices of German firms in North America. *Economic Geography, 81*(1), 31-52.

Irani, P., Slonowsky, D., & Shajahan, P. (2006). Human perception of structure in shaded space-filling visualizations. *Information Visualization, 5*(1), 47-61.

Senge, P.M. (1990). *The fifth discipline: The art and practice of the learning organization.* New York: Doubleday.

Tsoukas, H., & Vladimirou, E. (2001). What is organizational knowledge? *Journal of Management Studies, 38*(7), 973-993.

Wilson, T. (2002). The nonsense of knowledge management. *Information Research, 8*(1), Paper 144.

Zack, M. (1999). Managing codified knowledge. *Sloan Management Review, 40*(4), 45-58.

Endnotes

[1]   This is akin to theories of the fundamentals of matter, such as "string theory" which cannot be empirically measured.

[2] These dimensions may also be considered the basic attributes of knowledge management systems because they describe the functioning of these systems in organizational settings. See Dutta (1997), Senge (1990), and Zack (1999).

Chapter XIII

Epistemetrics:
Why We Measure

Every nugget of knowledge is relevant and useful. There is no knowledge that can be described as immaterial, irrelevant, unnecessary, or without potential use. Since knowledge advances and grows by means of cumulation, every nugget adds to the pool, like every brick which is an essential component of a wall. The only possible shortcoming of any nugget of knowledge is the extent to which the transactor or user of knowledge is able to cluster it with other nuggets in his possession. The fault in any knowledge not being considered relevant and useful is not in the knowledge itself, but in the transactor or the user (e.g., Card, 2000; Davenport & Volpel, 2001; Patriotta, 2003; Rajan, Lank, & Chapple, 1998).

Epistemetrics is the conceptual space in which we are measuring *what* we know, *how* we know, and *why* we know. This last topic of Part III consists of the outcomes from transactions in knowledge, and the impacts and benefits which accrue to the transactors and to others.

From Generation to Utilization

We measure knowledge because the actors who transact in knowledge (at the individual and organizational levels) gain from their transactions. Such gains or benefits accrued to them will provide the explanation of "why" we measure knowledge.[1] There are three types of transactors in knowledge: generators, transformers, and users. Each has distinct gains or benefits they derive from the pursuit and the transaction in knowledge.

In a way, all transactors are users and beneficiaries of knowledge. Generators benefit by more proximal outcomes, such as personal growth and competitive advantages. Transformers benefit from the contributions that knowledge provides to the workflow, the processes, and the activities of individuals and their organizations. Ultimate users benefit from most of the contributions of knowledge to generators and transformers and, in addition, they also benefit from outcomes that the use of knowledge seems to provide, such as economic, social, technical, and systemic outcomes.

Users of knowledge are individuals and organizations who implement, utilize, adopt, absorb, adapt, and exploit as well as benefit from the outcomes and impacts of knowledge. They incorporate knowledge into their activities; integrate knowledge with their skills, abilities, and competencies; and add it to their existing stock of what they know and understand of their environment.

At any given time individuals and their organizations may have multiple roles as generators, transformers, and users of knowledge. Consider an individual who generates knowledge and deposits it in a knowledge management system (KMS). The individual will do so because he is driven by a goal of personal growth, improved technical skills, and increased competitiveness. The individual believes that this goal can be achieved by the benefits that would accrue from the transaction in knowledge. By transacting in knowledge, the individual generates a variety of benefits from which individuals and organizations could be enriched, and these would be compelling factors that would drive other individuals to transact in knowledge.

Outcomes and Benefits from Knowledge

The key reason why we measure knowledge is the desire of users to gain benefits from the outcomes generated by their use of knowledge. There are five possible categories of benefits (perceived and actual) that may accrue from knowledge.

The benefits from knowledge contribute to the accomplishment of the goals that motivate people and organizations to transact in knowledge: to generate and share it with others. *Individual* benefits contribute to the stock of abilities and competencies of the individual. *Organizational* benefits contribute to improvements in operations, in processes (such as decision making and communication), and in strategic stance and ability to survive. Other benefits are *economic, social,* and *systemic,* where knowledge contributes to such variables as productivity, cost-cutting, quality, regulatory compliance, and ability to further disseminate and use additional knowledge.

The benefits from knowledge may be perceived or actual. Transactors in knowledge may desire certain benefits and believe that they have occurred.

Table 1. Illustrative benefits (perceived and actual) which may accrue to users and beneficiaries of knowledge in organizations

Category of Impacts/Benefits	Illustrative Impacts/Benefits
I. Individual/Human Resources Benefits	• Improved level of education & literacy (technical & general) • Improved individual competence • Improved level of motivation & satisfaction • Improved sense of empowerment • Improved communications, relationships, & use of KMS
II. Project/Work Group & Organizational Benefits (Processes & Proceeds)	• Improved efficiency of operations • Reduced level of resistance to change • Improved exchange of S&T knowledge • Harmonized & improved standards • Improved decision-making processes • Added unit & organizational credibility
III. Economic Benefits	• Increased productivity & time & cost savings • Improved growth & market share • Reduced barriers to innovation & trade • Improved rates of ideas generated • Improved competitiveness
IV. Social Benefits	• Improved capacity to meet changing national needs • Improvements in regulatory compliance and in safety, reliability, and quality of products and services • Improvements in health, transport, energy, and other social goods
V. System Benefits	• Higher rate of dissemination of knowledge • Overall value added to all users/beneficiaries

Actual benefits are those that can be measured. Improved technical skills, competencies, literacy, and satisfaction are examples of benefits from knowledge to the individual transactor which can be measured. Similarly, improvements in operations, productivity, and competitive position in the marketplace are also measurable contributions.

Utility and Value of Knowledge: Models and Metrics

The various categories of benefits from knowledge described in the previous section may generate value to the transactor in knowledge. Figure 1 shows a model of the generation of value from knowledge and its relationship to the goals of transactors and the outcomes from the knowledge in which they transact.

As the figure shows, value is derived through a process by which outcomes from knowledge are transformed into benefits, and these generate value. The model in Figure 1 offers a possible answer to the question: "How does knowledge contribute to value?" In this model we see the progression of outcomes-to-benefits-to-value. There are, however, situational factors and intervening variables which may affect the degree to which outcomes are generated from knowledge, then transformed into benefits and value.

These factors are shown in Table 2. They are classified as barriers or facilitators to the process of gaining value from knowledge. The variables shown are examples of such factors as culture and practice, ability to generate outcomes, and the use of benefits to create value.

Hansen, Mors, and Lovas (2005), for example, have identified other variables impacting the transfer of knowledge. Among the factors they list are: search and transfer costs, team tenure and size (for the teams that transfer knowledge in engineering projects), amount of knowledge obtained, absorptive capacity, tacitness of knowledge transferred, and providers' perceived competition.

But a question remains: "How are the outcomes from knowledge transformed into benefits and how do these in turn generate value to the transactors in knowledge?" As shown in Figure 1, a process of three stages is necessary to reach the point where outcomes may crystallize into value. A successful end to this journey rests upon the successful completion of all three stages. Consequently, the value created with knowledge depends upon the outcomes being transformed into benefits, *and* the benefits transformed into value, *and* the benefits being adopted by users so that value can thus be derived. The expected value of knowledge in this process will be the sum of the probabilities that all three stages will be successful, so that:

$$Pv \left(\begin{array}{c} \text{Probability} \\ \text{of Value} \\ \text{Creation} \end{array} \right) = \sum_{i=1}^{3} P_a \times P_b \times P_c$$

in which:

P_a = Probability that outcomes will be transformed into benefits.

P_b = Probability that benefits will be adopted and implemented by users/beneficiaries.

P_c = Probability that benefits will generate value.

As this process is completed successfully, the outcomes from knowledge are made into benefits. *Outcomes* from knowledge are the immediate results from the externalization of knowledge by the generators. When they transcribe

Figure 1. What transactors in knowledge want and what value is derived from knowledge

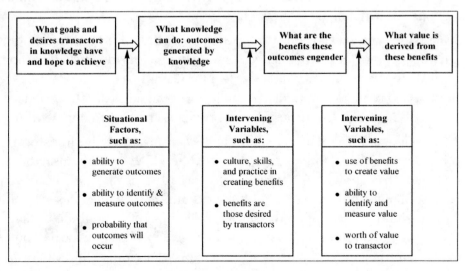

Table 2. Illustrative barriers and facilitators to the generation of value from knowledge

Factors Acting as Barriers	Factors Acting as Facilitators
• Lack of willingness to share what one knows • Sharing in untimely manner • Sharing and diffusing knowledge in a mode which is difficult to absorb by others • Previous negative experience • Sharing and diffusing marginal or irrelevant knowledge • Lack of willingness of others to receive or absorb knowledge • Benefits and ultimate value not recognized or perceived by knower • Cultural, organizational, and economic factors hindering sharing, and transformation of knowledge	• Perceived or recognized benefits • Competitive pressures • Perceived or recognized ultimate value • Past positive experience with benefits and value • Strong need for accomplishing goals (e.g., need for skills or improvements) • Cultural, social, economic, and organizational factors supporting sharing of knowledge and its transformations

the knowledge they possess and share it by communicating with others, this event generates measurable outcomes. These outcomes are usually in the form of intellectual "nuggets." They include concepts, ideas, formulae, lessons, conjectures, opinions, statements of fact, solutions, and answers to specific queries.

As shown in Figure 3, outcomes are generated when situational factors which can be barriers are overcome. In the next stage in this model, outcomes must be transformed into benefits. It is not enough for individuals and organizations to generate and to share ideas, lessons, or solutions. They must also incorporate them into their processes, procedures, and activities and make use of them so that they may derive benefits from these outcomes. Consider such benefits as "improved individual competence" and "improved decision-making process." Both benefits will be derived from the application and exploitation of such outcomes as lessons, ideas, solutions, and answers to questions.

The user of the outcomes (thus the beneficiary) must be able to overcome barriers and must have the ability to absorb and to utilize the outcomes so that they may generate the desired benefits. *Value* accrued to the transactor in knowledge is a concept different from benefit. In this chapter *value* and *utility* are used interchangeably. The concept connotes the ultimate overall contribution of knowledge to the transactor. Once benefits have been absorbed by the transactor in knowledge, the resulting contribution is defined as value accrued to the transactor. Generally, value may be the compilation of various benefits which may combine to engender gains defined as value.

Value may be generated in three possible forms. The first is in the form of *worth* with such content as prestige, financial and economic assets, social and ethical goods, and esteem. The second form is those *competencies, skills, capabilities,* or other *qualities* and *attributes* that the transactor in knowledge did not possess (or partially possessed) and has now acquired—thanks to the transaction in knowledge, its outcomes, and the implementation of its benefits. The third form includes things *intangible,* ethereal, or psychological, such as personal satisfaction, personal growth, feelings of accomplishment, achievement, and fulfillment. Strategically, outcomes and benefits may be viewed as instruments for the eventual creation of value so that what we want from knowledge and hope to achieve with its generation and sharing is ultimately the value we believe we may create at the end of the process of transformation.

Although benefits can be measured with relative ease and precision, value accrued from knowledge is far more subjective. Improved levels of skills, competence, productivity, or growth (examples of benefits) can be measured, whereas value such as worth, feelings, and sentiments are more difficult to assess and to measure. Yet, these intangible elements of value may be very important to transactors in knowledge. They usually represent a good portion of the goals and aspirations of transactors, and they explain why these individuals and their organizations would engage in producing, sharing, and diffusing their knowledge.

Illustrative Case

Consider the case of Mary who is a junior partner at a major management consulting company. Mary has been assigned to a large-scale project on automotive engines with a Chinese conglomerate. This was Mary's first assignment in China, and her first acquaintance with engine technology. Mary needed to acquire knowledge on both subjects. She contacted several colleagues, searched the Internet, and visited two libraries and two engineering departments of a prestigious university.

The outcomes generated by the knowledge Mary acquired included skills, general understanding of the phenomena of the manufacture and marketing of automotive engines, and the culture of doing business in China. The more Mary believed that the knowledge she would acquire would generate value for her successful completion of the project, the more she engaged her time and effort in the intensive search for and adoption of knowledge.

Mary was able to garner such knowledge in a relatively short time period only because she already had an extensive background of similar knowledge about business and technology, so the "new" knowledge could now be cumulated to the existing pool of knowledge in Mary's possession. "Continuous cumulation" enabled Mary to absorb the new nuggets of knowledge and to add them to her arsenal. The benefits for Mary extend beyond the successful completion of the immediate project. She is now endowed with improved individual competence, a strong sense of achievement, empowerment, and confidence. Her organization added credibility, improved performance, and gained valuable competitiveness in a very turbulent environment of global business.

Much of what Mary now knows is embedded in her mind and cannot be easily shared with others. Mary was able to garner additional knowledge by adding nuggets to her "personal" stock of knowledge about business, technology, project management, working globally, and similar topics. This effort had to be a personal journey. Mary could not have gained this knowledge by simply receiving it through communication from others. Mary, of course, had consulted with colleagues, but only to gain clarification and reinforcement to what she was already doing as her main effort. All of Mary's education and life and business experiences came to bear in this instance as she embarked on the search for knowledge.

Strategic Types of Value from Knowledge

The *expected value* from knowledge will be the sum of the probabilities that outcomes will be transformed into benefits and these in turn into value. In addition, these transformations depend on the degree of importance that beneficiaries from knowledge will assign to the expected value. This phenomenon is similar to the expectancy theory of motivation in organizations (Vroom, 1964).

$$EV \text{ (Expected Value)} = \sum_{i=1}^{n} Pn \bullet Im$$

where:

n = stages of transformation,

Pn = probabilities that outcomes and benefits will be transformed, and

Im = degree of importance beneficiaries attach to each value.

In the expectancy theory of motivation, performance of organizational members is the instrument used by individuals to attain rewards. Members will therefore be motivated to perform if they perceive the rewards that follow their performance to be of importance to them.

In the case of knowledge, transactors in knowledge will prefer the outcomes and benefits that will generate value of higher levels of importance. Transactors will therefore have more incentives to generate and to transform knowledge in which the outputs lead to those benefits which, in turn, accrue values preferred by the transactors. Consider an individual who generates knowledge and shares it with others (directly or by depositing it into a knowledge system). The individual is doing so because he is driven by the goals of improving his skills and competencies. These are the ultimate values this transactor would prefer to accrue. Moreover, the values will be inherently in concordance with what transactors desire and prefer, thus leading to a state of continuous stasis of knowledge generation and usage, as a self-organizing system (e.g., Shapiro, 2005; Leydesdorff, 2001).

The relationship between the values accrued to transactors in knowledge and the effort they expend to achieve these values is shown in Table 3. The higher the importance they perceive in these values, the more effort they will expend to generate and to share and transform items of knowledge with the highest probability of "surviving" the stages of outcomes and benefits.

Table 3. Strategic types of knowledge generated and diffused in organizations

H	(c) "Background"	(d) "Golden"
VALUES OR UTILITY FROM KNOWLEDGE	(a) "Immaterial"	(b) "Promising"
L		

Level of Effort Expended by Transactors in Knowledge

There are four strategic types of knowledge in organizations. Type (a) occurs when little effort is expended and the knowledge produced is of low utility or value. Such knowledge may be described as *immaterial.* For example, during discussions on a project with a Japanese company, a team member shares her knowledge of a recent project with a Korean company. Such knowledge did not require much effort to produce and has low utility at that time.

Type (b) is knowledge described as *promising.* This is knowledge produced with much effort but with low utility. Organizational members who expended such effort had certain values in mind. Whereas such values are not forthcoming at this time leaves them with a promise and hope that their effort will be justified.

A third type (c) is knowledge produced with little effort yet accrues high value. Such knowledge may be described as *background.* For example, individuals in the organization generate and diffuse what they know "off the top of their head" without expending much effort. Such knowledge may be of high value to other members and to the organization, primarily as background knowledge. This type of knowledge is generally needed in the organization to gain an understanding of circumstances, culture, procedures, and similar variables that describe business situations or the explanatory basis for how people and organizations make decisions. Such background knowledge is highly valuable in strategic monitoring of business competitors and in routine assessments of intelligence inputs from multiple sources (see Bryant, 2005; Hall & Paradice, 2005; Van der Penne & Dolfsma, 2003; Watts & Porter, 2003).

The fourth type of knowledge (d) is produced and diffused with much effort and is considered to accrue high value. This knowledge may be described as *golden*. Examples of such knowledge include the effort expended to generate and to share and diffuse knowledge which is not readily nor easily available on a specific, usually urgent topic critically needed for a current task or project. Knowledge gathered by intelligence agencies on planned terror attacks is such a type (d) or golden knowledge. Another example is the knowledge engendered and diffused by a business corporation on the launching of a highly promising new product by a competitor.

Clearly the measurement of goals, outcomes, benefits, and value is an activity which to be successful requires several prerequisites. First, there is a need to have all these variables defined and identified to an extent that measurement becomes feasible. Even variables that are subjective (such as perceived importance of benefits and perceived value) need to be adequately defined for them to be measured. Second, goals, outcomes, and benefits must not only be defined but also understood and utilized by transactors in knowledge as an integral part of their procedures and activities.

Clearly the measurement of goals, outcomes, benefits, and value is an activity which to be successful requires several prerequisites. First, there is a need to have all these variables defined and identified to an extent that measurement becomes feasible. Even variables that are subjective (such as perceived importance of benefits and perceived value) need to be adequately defined for them to be measured. Second, goals, outcomes, and benefits must not only be defined but also understood and utilized by transactors in knowledge as an integral part of their procedures and activities.

Linking Value from Knowledge to the "What" and "How" of Epistemetrics

Once knowledge is generated in the form of intellectual nuggets, these nuggets will be transferred, shared, and may be deposited in a knowledge repository for further diffusion and future use. What follows may be an array of processes in which knowledge is absorbed and utilized by individuals and organizations while generating benefits and ultimately value to those who transact in knowledge and who gain from its generation and use.

The road from generation of knowledge to the value it may accrue is a complex enterprise. It is replete with barriers and stages in the transformation of an intellectual nugget—from its generation to an identifiable and measurable value to the user of such intellectual nugget. If, for example, a company is about to engage in a project with a foreign collaborator, the project manager asks: "What do we know about this company x?" Any intellectual nuggets that will be generated by project members, those found in the company's knowledge repository, and those to be procured from other sources will have to undergo the transformations described in "How We Measure" knowledge. The value to be acquired from these nuggets, after their transformations, will be dependent upon the conditions, pre-requisites, barriers, and processes described in this chapter.

Key Issues in Why We Measure Knowledge

The first key issue is the complexity of the process—from generation to value. Multiple stages are embedded in the processes by which knowledge in the form of intellectual nuggets is transformed, shared, diffused, and utilized. In each stage there are multiple barriers which hinder the transition along the processes. This level of complexity makes it difficult to measure the transformations and the resulting entities from them (outcomes, benefits, and value) in a way that is adequate and acceptable.

Another key issue is the temporal and conceptual distances between the event of the generation of knowledge to the use of it and the accrual of value from it. This distance has the potential to create gaps in the timeliness, the readability, and the relevance of the knowledge being transformed. Different actors transact in an item of knowledge over periods of time. This may lead to knowledge that is no longer relevant to the topic at hand or no longer presented in a format that is readily understood by the downstream transactors.

A third key issue is the elusive nature of the value accrued from knowledge. In addition to its largely intangible nature, the value from knowledge is not easily configured or measured, principally because of the difficulty in telling *when* such value is actually being derived and to *whom*. For example, prestige and similar intangible values may become recognized within a certain timeframe, perhaps years after the value had been accrued. Similarly, even values such as competencies and capabilities may crystallize into recognizable and measurable value only after a substantial period of time and a host of unrelated activities.

A good example is the value accrued to students from knowledge they receive in their graduate-professional education. The perennial request of business students is for knowledge they can use without delay, hence knowledge that can accrue measurable benefits within days, perhaps even hours. When confronted with the alternative hypothesis that "better" knowledge would be that which will accrue value within a time period measured in years, not days, these students are often unconvinced and undeterred from their request for immediate production of value from the knowledge they had acquired. After years of on-the-job experience, many students (now executives) recognize their previous misconception and tend to agree that the more precious value was that which accrued to them from knowledge they later recognized and fully benefited from several years afterwards.

The models of the progression of knowledge from generation to utilization and the role of outcomes, benefits, and value have considerable implications for individuals and organizations who transact in knowledge. There are also lessons for the behavior of transactors and the choices they make.

For individuals, these models provide a connection between their goals and the ultimate value they wish to acquire from their transactions in knowledge. In order to accomplish their desired value, individuals may now better plan the processes of generation and diffusion of knowledge and the types of knowledge they will be more likely to generate and to diffuse.

An important implication for organizations is a better understanding of their strategies for the generation and diffusion of knowledge. These models allow for a more promising allocation of incentives for individual members to generate, transfer, and diffuse their knowledge. Organizations may also gain a better handle over the pace of the progression of knowledge by carefully intervening at specific points along the process of knowledge diffusion, through its outcomes and resulting benefits. Such interventions would be designed to improve the likelihood that outcomes would be transformed into benefits and that benefits would accrue value.

Organizations would also have additional insights into mentoring and training of individual members who transact in knowledge. By better understanding why individuals transact in knowledge and how value is created from it, organizations now have an improved capacity to target both mentoring and training to overcome people's resistance to share and diffuse knowledge. Similarly, training may now be an instrument to encourage and to promote the generation and diffusion of knowledge.

Organizations would have the ability to devise and to institute a knowledge-auditing system. Designed with the lessons from the models and the process of value creation, such an audit would assess the process of diffusion of knowledge, the barriers to such diffusion, and mechanisms that may serve to facilitate the generation of diffusion of knowledge. The audit would also evaluate the value accrued from knowledge and the role of such value creation on current and future transactions in knowledge in the organization.

We need to further explore the following questions. First, will more knowledge creation and sharing increase the amount and pace of benefits and ultimately of value thus accrued? Is there a possible effect of diminishing returns, so that beyond a certain point the value accrued from knowledge starts to decrease? Studies of information overload may provide some insights to the methodology that could be applied in the case of knowledge.

Another question that needs to be explored is the gap between individuals who transact in knowledge and their organizations. Why are some organizations more successful at utilizing the knowledge created and shared by their members? The reticence to share knowledge is a universal aspect of human behavior. Yet some organizations are better at establishing knowledge systems and at encouraging their members to deposit their knowledge into these systems and to share such knowledge with others.

We measure knowledge because individuals and organizations transact in knowledge under the belief and the hope that by doing so they will accrue desired benefits which in turn will create the value they want and cherish. To accomplish such benefits and resultant value, individuals are willing to overcome many barriers to the generation and the sharing of what they know. Organizations are also willing to establish mechanisms and specialized systems designed to encourage such behavior and to capture and utilize the knowledge thus produced and diffused. Epistemetrics is therefore instrumental in explaining what we measure in knowledge, how we measure, and finally, why we measure.

References

Bryant, S. (2005). The impact of peer mentoring on organizational knowledge creation and sharing. *Group & Organization Management, 30*(3), 319-339.

Card, J. (2000). Development and dissemination of an electronic library of social science data. *Social Science Computer Review, 18*(1), 82-87.

Davenport, T., & Volpel, S. (2001). The rise of knowledge towards attention management. *Journal of Knowledge Management, 9*(3), 162-171.

Hall, D., & Paradice, D. (2005). Philosophical foundations for a learning-oriented knowledge management system for decision support. *Decision Support Systems, 39*(3), 445-461.

Hansen, M., Mors, M., & Lovas, B. (2005). Knowledge sharing in organizations: Multiple networks, multiple phases. *Academy of Management Journal, 48*(5), 776-793.

Leydesdorff, L. (2001). *A sociological theory of communication: The self organization of the knowledge-based society.* Boca Raton, FL: Universal.

Patriotta, G. (2003). *Organizational knowledge in the making: How firms create, use, and institutionalize knowledge.* New York: Oxford University Press.

Rajan, A., Lank, E., & Chapple, K. (1998). *Good practice in knowledge creation and exchange.* Tunbridge Wells, UK: Create.

Shapiro, I. (2005). *The flight from reality in the human sciences.* Princeton, NJ: Princeton University Press.

Van der Penne, G., & Dolfsma, W. (2003). The odd role of proximity in knowledge relations: High tech in The Netherlands. *Journal of Economic and Social Geography, 94*(4), 451-460.

Watts, R., & Porter, A. (2003). R&D cluster quality measures and technology maturity. *Technological Forecasting and Social Change, 70*(2), 735-758.

Endnote

[1]   This utilitarian approach also takes into account the personal or emotional benefits to individuals from the activity of searching for knowledge and the pleasures derived from the mere pursuit of knowledge.

Section IV

Applications of Knowledge Systems in Organizations, The Economy, and Society

Chapter XVI

From the Garden of Eden to the Knowledge Society

The "Knowledge Race"

In the contemporary post-industrial society, we are continually creating and accumulating knowledge. There seems to be an unending quest to know more about less. Peter Drucker (1994) had a good explanation. He argued that in the past, workers in farming and even in factories were generalists, whereas knowledge workers are highly specialized. One of the crucial skills they develop is their ability to learn how to acquire additional specialized knowledge.

Once knowledge became the new source of wealth creation, society made the necessary rearrangements to accommodate the creation and usage of knowledge. This led to the proliferation of so-called "knowledge workers." Nowadays the organizations that employ these workers and to whom the workers market their knowledge assets are also continually restructuring

themselves to effectively deal with the continuous onslaught of the generation, manipulation, and utilization of knowledge.

The knowledge race and the modifications and restructuring of the institutions of the post-modern society are anchored in the viability and utility of
databases and knowledge warehouses and systems. These are the tools of the
new reality and the instruments by which knowledge can be manipulated.
Databases and knowledge systems—in electronic or other formats—are
the equivalent of the plough in historical agriculture and the manufacturing
machine in the industrial factory.

In order to "run" these new tools in ways that are both effective and efficient,
we are required today to understand how they are structured, how they work,
and also how to improve upon them. This need is so pervasive that whenever
we come up with an innovation in these areas, the payoff to the inventors is
beyond their wildest expectations.[1] In the race to gather more knowledge of
better utility and in a format that conforms to the tools at our disposal, we find
ourselves in a game of "chicken or egg." Databases and knowledge systems
are continually updated to "fit" the stream of knowledge being accumulated,
whereas the characteristics of the knowledge being sought are matched to the
capabilities and the attributes of the databases and knowledge systems.

The knowledge race creates a dramatic transformation in contemporary life
as influential as that of the industrial revolution and, before that, the collapse of feudalism and the beginning of urbanization. Just as writers and
philosophers in the nineteenth century lamented the demise of the bucolic
agricultural society and the emergence in its turn of the industrial way of life
(unforgiving, brutal, and terribly demanding on body and soul), so we are
beginning to see today the lament upon the knowledge society. Again, we
have left the "Garden of Eden" to embark on a merciless journey of incessant
quest for knowledge that is transforming us into merchants of what we know
and makes us pitiful victims of unrelenting obsolescence.

Interdependency and the "Colony Effect" of Knowledge Cumulation

Peter Drucker was correct when he argued that specialized knowledge requires
knowledge workers to work in teams. We are now dependent upon each other
because none of us has all the knowledge needed to carry out tasks and activities in the knowledge society. Knowledge work has become a jigsaw puzzle.

To complete it one needs the interdependent parts to work together.

The power of knowledge lies in the collective assembly of all specialized knowledge. This pattern of work can be described as "The Colony Effect."

In the post-industrial society, we continually accumulate knowledge, particularly because of specialization and the need to extract as much utility as possible from such knowledge by means of the collective cumulation. A good analogy would be the way colonies of insects share their knowledge.

In societies of insects, such as honey bees, there is specialization in terms of task partitioning so that the members of the colony transfer materials to one another in a good example of teamwork and division of labor within the team.[2] Honeybees also exchange knowledge on such crucial topics as location of fertile and usable areas to forage for food and building materials. The knowledge base of the colony is the totality of inputs from individual foragers. It depends on continuous and varied inputs due to changes in the environment and other threats that impact locations and their usability.

Human beings in the contemporary age collect vast amounts of knowledge within the framework of their organizations. Government entities and industrial companies assemble knowledge from a variety of specialties and functions. By themselves each specialized item of knowledge has limited value to the collective. But when massive and diverse knowledge is accumulated, the potential value to the task at hand and to the organization and society increases drastically—providing that the people, their organizations, and society at large can process this mass of knowledge.[3]

Here is the rub! In order to perform collective tasks such as organizational production or strategic goals of public institutions such as defense, post-industrial organizations must absorb and process knowledge so as to derive value from it for the effective discharge of their tasks. This is a tall order. Its successful completion depends on their abilities to generate adequate specialized knowledge (with promise of utility), and to cluster and interpret such knowledge in a manner that will benefit the larger organization or community—across specialties and disciplines.

Notice that I used the verb "to cluster" as a reminder of how the human mind deals with the flow of sensorial inputs in order to generate knowledge. The obvious implication, which I discuss in Chapter XV, is that to fully exploit the "colony effect" of knowledge in the post-industrial society, we should build mechanisms that process knowledge and emulate or at the very least resemble the brain's mode of structuring knowledge.

The Value of Knowledge in Post-Industrial Society

As we accumulate and process knowledge, how do we know that we are indeed deriving value from this extensive effort? Gaining and adding knowledge in itself may not contribute adequately to the addition of *useful* or valuable knowledge, meaning: adding knowledge but not "knowing" more—quantity but not quality.

This phenomenon may be called "The Insufficiency Effect." We add knowledge but it adds little or no value to our stock of what we know. This is quite possible, assuming that we can measure the value of knowledge and that we can adequately define what constitutes value of the knowledge we add to our stock.

What is the value generated by knowledge? I will use here the analogy of the value we attribute to R&D (research & development), and science and technology. In my books, *The Metrics of Science and Technology* (2000) and *Creating Value with Science and Technology* (2001), I held the notion that value can only be defined indirectly by the benefits and contributions that can be attributed to science and technology. The same applies in the case of knowledge. Value derived from knowledge is the sum of the benefits and contributions that knowledge bestows on the knower (self) and on others. As in the case of science and technology, we define the value of knowledge by its benefits, which can be attributed to specific recipients (such as individuals and organizations), hence allowing us to develop some measures of what value is.[4]

But, value to whom? Who are the recipients of the benefits from knowledge, and how do we measure these benefits? Are there differences between benefits to the knower and to others?

Measuring the Value from Knowledge

To measure the value derived from knowledge, we must establish to whom the value is accrued. Clearly, value can be accrued to the knower (self) and to others (the collective, the organization, a community, or society). Table 1 shows the classification of types of benefits from knowledge to the two types of recipients: self and others.

Benefits from knowledge are not measured in the way we measure, for example, electricity (KW), sporting events, or other tangible activities with outcomes. We measure knowledge *indirectly,* by the benefits it generates and the contributions it makes to the knower and to others. As in the case of science and technology, there are more proximal benefits, such as those to the knower, and benefits which accrue down the line at the organizational and social environments.

The measurement of the benefits from knowledge (as from science and technology) is a subjective assessment of what knowledge seems to contribute to the knower and to others. We can therefore merely estimate the impacts of knowledge. The recurrent definition of a benefit is in terms of "improvement." We subjectively evaluate the benefits from knowledge, but we are yet unable to quantitatively measure the degree to which such improvements indeed took place.

A good estimation of some degree of assessment is shown in Figure 1. The range of value created by knowledge runs the gamut from marginal to indispensable. In situations of danger to the knower or the collective (such as "fight or flight"), the benefit from knowledge about the danger and how to deal with it may be indispensable.

The value of *added* knowledge in grave situations is high, but its contribution is not limited to the "item of knowledge we had just added." Rather, the model of the progress of knowledge credits the catalog of knowledge as a contributor to the benefit. For example, in a situation of "fight or flight," the knowledge about the latest direction from which the threatening force is coming would be indispensable to a good decision on which action one should take to deal with the threat. However, in order to utilize this item of knowledge, one must have a pool of knowledge into which this latest item can be incorporated, so that the decision will be based on clustering this "new" knowledge with existing knowledge about alternative actions, past

Figure 1. A continuum of levels of benefits from knowledge

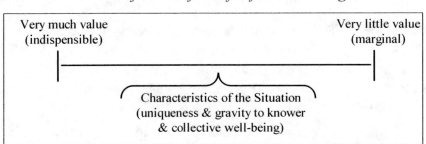

Table 1. A typology of benefits from knowledge

Type of Benefit	To Knower (Self)	To Others (Collective)
A. For the sake of knowledge	*Examples of benefits:* • Satisfaction • Wisdom • Understanding	• cultured members • improved civilized attitudes and behavior
B. Adding to the stock of knowledge	*Examples of benefits:* • Improved understanding • Improved ability to make decisions	• improved performance • improved ability to survive • improved ability to change
C. Means to such aims as economic and social goals and activities	*Examples of benefits:* • Gaining wealth • Better chance to survive and succeed • Better chance to perform in collective	• improved attainment of economic goals • improved attainment of social goals • improved ability to grow
D. Embedded in other things, such as products and services	*Examples of benefits:* • Better performance of products and services • Improved human-machine interaction	• technological and scientific breakthroughs • improved economic and social sophistication

experience with such threats, and knowing the strengths and limitations of the decision maker.

By integrating the value of added knowledge with the existing pool, I am also proposing that the relative value of a given item of knowledge does not necessarily increase or decrease as the knowledge stock grows. The chances are better for the relative value of added knowledge to *increase* as the knowledge pool increases. Simply put: the more we know (in general and on a given topic or subject), the better we can cluster, integrate, and utilize added items of knowledge, thus improving the possibility that such items of knowledge will add value.

Key Problems with Metrics of Knowledge

I am thinking of two key problems. The first is: Can we adequately *quantify the value* accrued from knowledge? How do we quantify knowledge and how then do we measure the benefits? This is hardly a simple problem, and although I have devoted much of my intellectual pursuit to problems of measurement of unstructured phenomena, I find it difficult to satisfactorily quantify these benefits. The combination of the subjective nature of the evaluation and the general nature of the benefits described in terms such as "improvements" makes this a hard task to accomplish (e.g., Geisler, 1999).

Another problem is the arduous task to *isolate the effects* or impacts of knowledge, so as to establish benefits to social and economic systems. This also applies to the benefits from knowledge that are embedded in products and services. How do we isolate the impacts of knowledge vs. other causes and factors? Again, as in our effort to quantify such benefits, we subjectively assess the impacts and estimate the role that knowledge played in creating these benefits.[5]

Based on the benefits that knowledge generates, it may even be possible to define knowledge as a *useful* effort. Knowledge not only provides us with an understanding of our environment, but it also generates utility in the form of benefits. This unique attribute may be another factor that distinguishes knowledge from data and information.

Yet, however we define and measure value from knowledge, the fact remains that it would be impossible for individuals and collectives to survive, to grow, even to function—without knowledge. In the post-industrial era when so much depends on knowledge, we can safely assume that knowledge does create benefits—albeit we cannot yet accurately measure them beyond estimates of possible impacts.

Hence, all knowledge we acquire has some value, extending beyond our effort to estimate its utility in a particularly immediate situation. All knowledge has value that is also a function of the pool of knowledge already in existence with the knower or the collective. As the continuous cumulation model proposes, we cannot assess the value of knowledge unless in conjunction with the body of knowledge that supports it and of which it is an integral component.

A Personal Note on Value from Knowledge: Are We Really Banished from Eden?

There is a strong current of people who have expressed their discontent, even pessimism and fear, at the unhindered growth of knowledge. The often-used term is "technology" and the perils its growth brings to our lives.

This perspective can seamlessly be extended to knowledge. Albert Einstein expressed his concern that humanity is not yet ready to handle the level of knowledge and technology that scientists are bestowing on it. More recently, Bill Joy, co-founder of Sun Microsystems, lamented that nanotechnologies would become so embedded in our lives that we would lose control over this knowledge and these technologies (Joy, 2000).

I hold a more optimistic view. As we continually accumulate knowledge (and some of its advanced forms becomes embedded in aspects of our lives), we also increase the probability that we would know how to control it. By opening this "Pandora's Box" of dangerous knowledge, we also generate knowledge *about* the perils and ways and means to confront these challenges. It is the power of cumulation of knowledge that offers the best hope for the future.[6]

Once upon a time we dwelt in the Garden of Eden, wandering happily, unaware of what was brewing in our surroundings and in our future. This was the time when we played with databases and we unleashed the initial power and impact of digital computing. We kept accumulating volumes of data and we conducted analyses of their characteristics. We reveled in the wonder of statistics and shared a vivid belief in the bounty of data and databases.

Then, along came the era of information. We awoke to the promise of information systems, management information systems, and the impacts of information in our lives. Suddenly our world was transformed. In the latter years of the twentieth century, we entered a period of what was called "The Information Age."[7] James Dewar of the Rand Corporation wrote an insightful analysis of the information revolution, by comparing it with the invention of the printing press (Dewar, 1998). He argued that there are "provocative parallels" between the impacts of the printing press and networked computers on communication and the diffusion of information, particularly exchanges of scientific information (and knowledge). The parallel with the printing press also leads to predictions that networked computers—with the main example being the Internet—will generate changes in social and economic trends, and generally in the way we live our lives for many years to come.

The arrival of the era of information was due to the plethora of questions we began to ask of our mounds of data. Suddenly, it was not enough to have many data-points on a chart. We started asking what these data-points represent and how they can help us define and solve problems. As organizations saw their spendings on databases continue to sharply increase, their owners and managers started asking incisive questions about the value derived from these investments. Adding to this pattern was also the move, in many organizations, from backroom data processing to the "front office." New software now allowed managers to apply information (in the form of analyses of large quantities of data) to sales, marketing projections, and strategic planning.

This is, very briefly, the story of how we came to possess this tool of information systems and its extension as management information systems (MISs). But, this was not enough. Our departure from the gullible existence in the Garden of Eden also meant that we began to entertain notions of knowledge that information may be generating for the organizations that find such systems.

In the decades from the mid-1990s to the publication of this book, we witnessed the emergence of the new movement of knowledge management. Self-appointed "experts" appeared on this hastily constructed stage. Some of these new "gurus" exported their wares from their work in information analysis and information systems. Others arrived from the organizational, managerial, and communications arenas. I must confess that my own academic trajectory also catapulted me straight into the midst of this activity.[8]

In this chapter I introduce a critique of the existing systems of knowledge in our organizations. I will show that in moving from the information society to the knowledge society, we created a reality that is at best imperfect and mostly containing promises that cannot be kept. In this chapter I also lay the ground for the *thesis of disconnect,* and for my suggestions how to improve on our fragile grasp of knowledge and its utilization.

What's Wrong with Knowledge Management?

In the early years of the twenty-first century, there was much jubilation in the knowledge management movement. The intellectual activity had produced a growing number of papers in the various literatures related to this topic. An analysis of these papers in 2003 suggested that in the period 1990-2002 there were no fewer than eight different subfields of research (Subramani, Nerur, & Mahapatra, 2003). Yet, the prolific output of papers had failed to develop a focus or to generate significant contributions.[9]

A harsher critique comes from Europe. Professor T.D. Wilson of the University of Sheffield in the United Kingdom argued that knowledge management is a fad, constructed and publicized by management consultants (Wilson, 2002).

Wilson based his conclusion on what he believes to be the organizational foundation of knowledge management: a culture that encourages and allows for unhindered sharing of knowledge. He also passionately criticized the notion embedded in knowledge management that "tacit" knowledge can be captured and transferred into explicit knowledge. By focusing on the work of Nonaka and Takeuchi (1995), Wilson suggested that their use of the term "tacit" knowledge was incorrect and should have been replaced with the term "implicit" knowledge. The example of the bread-maker (given by Nonaka and Takeuchi) used by Wilson to explain his critique leads me to believe that he was referring to *procedural* knowledge (how to make bread).[10]

In the larger picture of knowledge management in organizations, these examples of sharing procedural knowledge are simply the essence of *mentorship.* In pre-industrial times the mentor transferred the "tools of the trade" to an apprentice. This practice continued to a degree during the industrial age. In the post-industrial era, there is less and less of this venerable tradition, so precious procedural knowledge and organizational experience are lost when experienced workers depart the organization. Due to ever-increasing specialization and the hectic competitive environment in today's organizations, managers and professionals are weary of divulging their "tricks of the trade" and their knowledge of the political and social processes that helped to elevate them to their position and to keep them there.

I agree with Wilson's basic argument that the early scholars (and consultants) in knowledge management misunderstood the notion of "tacit" knowledge as it was originally defined by Polanyi (1958). With the knowledge management tools available to us today, we are still unable to capture, translate, and share tacit knowledge.

But my critique of the current capabilities and promises of knowledge management is much more profound and goes to the core of what the structure of knowledge is. Wilson forcefully argued that knowledge management is merely "an umbrella term for a variety of organizational activities, none of which are concerned with the management of knowledge." These are *organizational* issues that can be addressed and perhaps even resolved by organizational means (e.g., Rubenstein & Geisler, 2003).

When we left the Garden of Eden—of just having data processing—we found ourselves in an imperfect world of knowledge management. This is a complicated and complex world where some very pesky questions and lofty expectations are constantly hurled at us. We have not fully or really moved from managing data and information to managing knowledge. In my view, we are no longer in Eden, but we behave as if we were still dwelling in the wonderful world of knowledge systems that can answer questions and impact our lives.

Let's face it: we are swimming in an ever-expanding ocean of data or information, and we have very few and effective tools to make sense of it. To better explain my view, examine the stages or levels of accomplishment of knowledge tools that are listed in Table 2. The table shows three levels of capturing knowledge.[11] At present, we are only at the level of *access* in which we can simply determine the position of a word or phrase in the database, and some of its statistical attributes.

Consider the example of the largest database of all: the Internet. Even the more advanced search engines are capable of merely providing access to the Internet. The use of Boolean logic or search techniques is based on three operators: OR, AND, NOT. For example, a search may look for "Clinton AND Bush" and locate all references or Web sites in which the two names appear together. Another option is the command NEAR in which terms will be found in proximity to other terms (such as within 10 or 25 words of each other). Compound searches may use more than one Boolean command—for instance: "New York AND Senate Race (Clinton OR Schumer)."

In all the examples we find ourselves only at the level of access. We determine where these terms are found, but not their *content* or their *utility.* This brings to mind the analogy with healthcare delivery. *Access* means that you are admitted to a hospital or a clinic, but it does not mean that you are privy to the diagnosis, the quality of care, the length of stay, the cost of treatment, and the prognosis for cure and survival. In other words, access does not generate knowledge.

Table 2. Levels of capturing knowledge from large data and information bases

Level	Description
1. Access	• Determining Where A Given Word, Phrase, Or Notion Exists In The Database • Determining Frequency And Other Statistics
2. Content	• Determining The Meaning Of The Phase Or Notion • Establishing The Knowledge Embedded In The Item
3. Utility	• Determining How This Item Can Be Applied To Solve Problems • Determining How This Item Contributes To The Knower And The Organization

I am reminded here of a scene from the movie "Midway." When told of a sighting of enemy (Japanese) ships and their positioning near the Island of Midway, Henry Fonda, in the role of Admiral Chester Nimitz, irritably snaps: "What kind of ships? What direction? What speed?" The mere position of the items is hardly meaningful nor useful knowledge. In the current state of search engines, we cannot ask the questions that will allow us to capture the knowledge embedded in the Internet.

The Double Curse of Semantics and Semiotics

In my opinion, the problem with the current state of knowledge management is deeper than the level of technological sophistication of our search engines and the organizational barriers to implementation of knowledge management systems (KMSs). In the past several years, we heard the term "disconnect" used and abused in a variety of areas and analyses. Nonetheless, I will use it here to describe the impacts of our clinging to the notions of semantics and semiotics.

The main problem with knowledge management today is the reliance on modes of gathering and sharing information that are poorly adaptable to the special needs of knowledge. I call this phenomenon: the *Thesis of Disconnect.* Simply put, there is a wide gap, or disconnect, between the modes we use to search, retrieve, analyze, and construe items of knowledge—and the modes in which knowledge is structured and progresses.

To extract and generate knowledge from databases, we use a combination of semantics and semiotics. These were modes that we used very effectively throughout the information revolution of the twentieth century. I will briefly remind the reader what they are, and why they are sadly inadequate in the age of knowledge.[12] Both semantics and semiotics are focused on communication, sharing, and transfer mechanisms.

Semantics explores the use of words and language in human interaction.[13] In a long line of studies and analysis of semantics, two influential scholars come to mind. They addressed two key questions: (1) what is the relationship between semantics and reality, and (2) what is the relationship between semantics and the human mind/brain?

Rudolf Carnap (1891-1970) had argued for "tolerance" in the use of semantics.[14] He suggested that logical analysis can be formulated by the use of syntax, so people have the freedom to construct reality and logical reasoning by their individual choices of words and syntax. In an analysis of the use of language to describe abstract entities, Carnap also stated that the relationship between language and universal or abstract entities is a matter of degree. In his framework, language describes abstract entities simply because it is a useful and expedient way to do so. We should not conclude from this, Carnap explained, that abstract concepts can be derived by logic rather than through empirical experience. The fact that we can use words to describe abstract concepts cannot be used to justify or support either school of philosophy of knowledge.

Language describes reality and words are assigned to specific entities, events, and constructs. Semanticists have explored the variations in the use of words, syntax, and other attributes and combinations of words. In time they separated the discourse from that of philosophers by avoiding the issue of ontology, that is, how real is the world we describe with words.

It was up to Noam Chomsky to explore the second question: how language relates to the human brain and to human cognition. He argued that language needs to be analyzed and understood within the context or circumstances in which it is used, and the person who uses it. He also advanced the notion that language is a faculty of the human brain and that humans are biologically endowed with some universal principles of grammar, that is, a scheme shared by humans that allows them a common platform by which they are able to communicate. Such natural language is constrained by the biological characteristics of the human brain.

Semiotics in its simplest definition is a study and use of signs as a form of communication and human interaction. The term "signs" also includes words, images, gestures, and social conventions such as myths, objects, routines, and symbols. Semiotics can also be the use of pictures by which people communicate and are able to assign a common meaning to the picture.[15]

In their large variety of definitions, signs are used to interpret social and anthropological events and phenomena. They help analysts to explain how people interact. Roughly, semantics looks at the meaning of words, whereas semiotics also deals with how the meaning of signs is woven into the social situation. For instance, a person may say the words "religious ritual" which has a meaning that other people can understand, but how religious rituals are used to maintain social cohesion or to teach the next generation—this is where semiotics will take over and attempt to explain these social aims via the symbol of "religious ritual."

So, why the double curse of these apparently useful means of analysis of human communication? Precisely because we use words and symbols as the preferred (perhaps only) means to communicate and to interact as only human beings can do so well. We have existed under the false belief that we can exchange and share *knowledge* with words and symbols. We surf the Internet and extract its contents by the use of *keywords*. This is a very primitive way of extraction from a database. It is analogous to looking up a telephone number in a directory. In a way it resembles the old adage about banks which only lend money to those who prove they do not need it. The primitive mode of accessing databases is anchored in the adage that you can find what you are looking for, if you know what you are looking for!

Besides the use of keywords, we are also cursed by semiotics.[16] Scholars in this field are debating whether signs are an instrument for social interface, or a construct of the human brain, or both. Another issue is whether there is a correspondence between concepts and the words we use to describe them, and if such signs (our mental pictures) that I possess are the same or similar to those that you possess.

There is, however, agreement among scholars that signs and symbols are arbitrary constructs based on cultural conventions. As communications experts will undoubtedly concur, the use of certain images differ by culture. In one set of cultural values, a sign or image may have a powerful message and a strong connotation, whereas in another culture it may just be of marginal interest without evoking any emotional response.

So, even when we create and store images and signs in our databases, and even when we retrieve them and attempt to make sense of them, we fail to extract their meaning and share their value as items of knowledge. Signs and symbols, including pictorial semiotics (mental pictures), are mainly an instrument for social exchanges. We cannot obtain from them answers to questions we wish to pose—as would be necessary if we wish to extract knowledge from them in the databases we are scanning for knowledge.

How We Search is not how We Know

The double curse of semantics and semiotics is upon us because those two modes of searching databases (or information bases) do not allow us to cluster. They simply give us a description of what is in the database in terms of words and images. But this is not how we structure knowledge.

Somewhere along the evolutionary process, humans crossed the barrier between being a repository of data to becoming processors of knowledge. The human brain was now able not only to store sensory inputs but to cluster them, so as to form higher-order constructs and to increase the stock of knowledge. Unlike lower-level organisms (such as bacteria) and the genetic code—where changes are only possible through mutations and transfer to multiple generations—humans could now process knowledge about their environment *independently of their genetic makeup and its constraints.* They were now able to influence their environment and to dramatically extend their capabilities and their reach by knowledge-laden science and technology.

Yet, our databases are still devoted to words and symbols. We cannot "ask" them questions that require clustering and the creation of knowledge which is embedded in them. By creating these databases in the post-industrial society, we lost our native ability to communicate what we know. We designed and built data and information systems whose principles of structure and access are divorced from the way we structure knowledge and the way knowledge progresses.

We have created what I would call *Model A*: a model of *primitive* handling of data and information. This primitive model is prevalent in most aspects of our lives. I will illustrate its impacts with two examples: gathering and analysis of intelligence, and industrial and organizational performance.

The Illustration of the National Intelligence Apparatus

After the horrible event of September 11, 2001, the United States had begun a radical reexamination of its national intelligence gathering and analysis apparatus. This culminated with the final report issued in August 2004 by the National Commission on Terrorist Attacks upon the United States. Among its many findings, the commission cited problems within the intelligence community that needed to be addressed. The recommendations for corrective action included: (1) "unifying the intelligence community with a new National Intelligence Director," (2) "unifying strategic intelligence and operational planning…with a National Counterterrorism Center," and (3) "unifying the many participants in the counterterrorism effort and their knowledge in a network-based information sharing system."

As an *organizational* solution, such concentration of effort in a centralized unit with powers and responsibilities to discharge counterterrorism is a worthwhile approach. But, as a solution to the generation of credible and useful *knowledge,* this approach is not enough. As long as the information bases and the modes of extraction of their content remain as they are, no amount of networking or sharing will generate the *knowledge* that we really need.

A good analogy would be to consider the intelligence community as a human body. Imagine that each function of our body has its own little brain attached to it. The digestive, respiratory, and vascular systems would each have its own brain, performing tasks of control, maintenance, and even improvement in performance. This is how the national intelligence apparatus functioned before 2001.

Now consider the commission's recommendation. It would be like gathering all the little brains in one location, with all of them inter-linked and cooperating. This is not how the human brain works. The human brain is highly centralized, where it is departmentalized by function, not by location. Control tasks are concentrated in one location and apply to several bodily functions. The human brain thus has the capacity to cluster the sensorial inputs it receives. The new organizational entity recommended by the commission would be simply a repository for the inputs from the various little brains. Unless it has the capacity to cluster and to answer questions that have answers embedded in the inputs, it will not generate knowledge. Rather, all it will generate will therefore be regurgitated data or information.

The Illustration of Corporate Performance

Consider the example of a manufacturing company producing a product made of metal. I consulted for such a company in an attempt to improve its research and development activity. My interviews with the sales force revealed that for several months salespeople had been providing their managers with information about some tests undertaken by small entrepreneurial companies in which a plastic substituted for the metallic product with very promising results. The company managers dismissed the information. They did not or could not cluster this input with reports from their research center that such substitution was technically feasible, nor with inputs from their customers that budget considerations may become powerful incentives for them to switch from metal to plastic.

The availability of data and information, and even the *sharing* of such data from a variety of different sources (in a "networking mode") is not at all a guarantee that clustering will occur. In fact, such inputs are incorporated into databases and information warehouses where they are structured and accessed by keywords and signs or images—yet without the clustering mode with which the human brain operates so effectively.

Knowledge Systems and the Primitive Model A

We can now conclude that our knowledge systems are currently built on information and data systems. They are structured and accessed by Model A in a primitive and wholly ineffective fashion. Our current knowledge systems do not produce knowledge. They are very different from the way our brain structures knowledge and the way knowledge progresses.

In the next chapter I will introduce the *Neuronal Model* (Model B) and argue for it as an attempt to utilize our understanding of how knowledge is structured in the design and use of knowledge systems. We must cross the chasm of the great disconnect between the way we handle knowledge in our data and information bases, and the way knowledge is structured and progresses.

References

Dewar, J. (1998). *The information age and the printing press: Looking backward to see ahead.* Rand Corporation Working Paper.

Drucker, P. (1994). *Post-capitalist society.* New York: Harper-Collins.

Geisler, E. (1999). Harnessing the value of experience in the knowledge-driven firm. *Business Horizons,* (May-June).

Geisler, E. (2000). *The metrics of science and technology.* Westport, CT: Greenwood.

Geisler, E. (2001). *Creating value with science and technology.* Westport, CT: Greenwood.

Joy, W. (2000). Why the future doesn't need us. *Wired Magazine,* (April), 237-262.

National Commission on Terrorist Attacks upon the United States. (2004). Executive summary (pp. 20-21). *9/11 Report.* Washington, DC: Author.

Nonaka, I., & Takeuchi, H. (1995). *The knowledge-creating company: How Japanese companies create the dynamics of innovation.* Oxford, UK: Oxford University Press.

Polanyi, M. (1958). *Personal knowledge.* Chicago: University of Chicago Press.

Rubenstein, A.H., & Geisler, E. (2003). *Installing and managing workable knowledge management systems.* Westport, CT: Praeger.

Subramani, M., Nerur, S., & Mahapatra, R. (2003). *Examining the intellectual structure of knowledge management, 1990-2002—an author co-citation analysis.* MISRC Working Paper 03-23, University of Minnesota, USA.

Wilson, T. (2002). The nonsense of knowledge management. *Information Research, 8*(1), 1-56.

Endnotes

[1]   A good illustrator is the Internet. The developers of the search engine Google became instant billionaires. Another example is the innovation in software, in which Bill Gates and Microsoft created a workable operating system for personal computers, with compensation that helped establish one of the richest and more successful companies in history.

[2]   See, for example, Anderson, C., & Ratnicks, F. (1999). Task partitioning in insect societies: Effect of colony size on queuing delay and colony ergonomic efficiency. *The American Naturalist, 154*(5), 521-535. Also see the work by Beekman and her colleagues in Australia: Beekman, M. et al. (2004). Comparing foraging behavior of small and large honey bee colonies by decoding waggle dances male by foragers. *Functional Euology, 18*(3), 829-835.

[3]   Insects generate and exchange knowledge geared for specific tasks targeted for survival of the community. See, for example, Seeley, T. (1995). *The wisdom of the hive.* Cambridge, MA: Harvard University Press; and Frisch, K. (1967). *The dance language and orientation of bees.* Cambridge, MA: Harvard University Press. This is a fascinating

study of how bees communicate and exchange knowledge by performing a dance whose movements are meaningful allegories of flight direction, that is, maps and patterns of suitable patches of resources and how to get there.

4   In the case of science and technology, I based the model of valuation on the *outcomes* from science, technology, and R&D. The focus of the model is the process by which R&D progresses within the organization in which it is generated and is then transformed along the way to create outcomes that impact the focal organization and others. In the case of knowledge, the state of our methodology is not yet developed to the extent that we can adequately measure the process of knowledge within organizations—as an organizational activity. Rather, we are limited to a model (shown in Table 1) in which we identify categories of benefits from knowledge.

5   The example of R&D and science and technology is also applicable here. See Geisler (2001).

6   The American public holds a similar opinion. The National Science Foundation conducts periodic surveys of the public's views on science and the wisdom of continuing funding of it. Consistently the people's view is a resounding support for public funding. They identify and acclaim benefits in agriculture, transportation, telecommunications, defense, and healthcare.

7   The literature on the "information age" is so vast that I will not even attempt to summarize it nor select the best readings. Let me, nonetheless, suggest a few illustrative publications that may help to illuminate the nature and impacts of the information age: Toffler, A. (1992). *The third wave.* New York: Bantam Books; Dertouzos, M. (1998). *What will be: How the world of information will change our lives.* San Francisco, Harper; Essinger, J. (2004). *Jacquard's web: How a hand-loom led to the birth of the information age.* New York: Oxford University Press; Berkowitz, B., & Goodman, A. (2002). *Best truth: Intelligence in the information age.* New York: The Free Press; and Applegate, L., Austin, R., & McFarlan, F. (2001). *Creating business advantage in the information age.* New York: McGraw-Hill.

8   I produced several papers and a book with a colleague. See Rubenstein and Geisler (2003).

9   I have paraphrased the study's conclusions.

10   I agree, in principle, with Wilson that tacit knowledge cannot be easily nor readily transferred.

11   Some consultants have developed frameworks with four levels: awareness, access, applications, and perception of utility and effectiveness.

12   This discussion is not about the differences between *information* and *knowledge*. As I consistently argued earlier in this book, I define knowledge as a process of clustering of sensorial inputs, whereas data and information are descriptors of events and entities in the environment. They do not constitute knowledge unless and until their incorporation into the process of clustering.

13   From the extensive literature, see the following examples: Kearns, K. (2000). *Semantics*. New York: Palgrave McMillan; Chomsky, N. (2002). *On nature and language*. New York: Cambridge University Press; Carnap, R. (2003). *The logical syntax of language*. Peru, IL: Open Court; and Carnap, P. (1988). *Meaning and necessity: A study in semantics and modal logic* (2nd ed.). Chicago: University of Chicago Press. Also see: Lappin, S. (1997). *The handbook of contemporary semantic theory*. Oxford, UK: Blackwell.

14   Carnap's "Principle of Tolerance" (1988) offers a wide latitude for people to use language and to make sense in what they communicate. He tells a story that this view of language was revealed to him in a dream or vision during an episode of illness.

15   See, for example, the original work of Ferdinand de Saussune in: Thibault, P. (1996). *Re-reading Saussune: The dynamics of signs in social life*. London: Routledge; and: Harris, R., & Kemp, D. (2004). *Saussune and his interpreters* (2nd ed.). Edinburgh: Edinburgh University Press. Also see: Pink, D.A. (2005). *Whole new mind: Moving from the information age to the conceptual age*. New York: Riverhead.

16   The reader may find a different "semiology" used in some books on the subject. Saussune had coined the term *semiology,* provenient from the Greek word for "sign." American scholars prefer the term *semiotics,* coined by Charles Sanders Peirce (1839-1914), an American philosopher who founded in the United States the school of *pragmatism.* Peirce contributed to our thinking about the scientific method. He introduced the notion of *abduction,* as an added mode of logical analysis of *induction* and *deduction.* Simply put, abduction (which he also termed

"hypothesis" or "retroduction") is a form of logical analysis in which an analogy produces conclusions beyond the properties of things. For example, one may argue about the location of a sample or a population from other properties such as size. For further reading, see Peirce, C. (1998). *Chance, love, and logic* (reprint ed.). Lincoln, NE: University of Nebraska Press.

Chapter XV

The Neuronal Model of Knowledge Systems, Data Mining, and the Performance of Organizations

Why We Cling to the Primitive Model

In the previous chapter I introduced Model A, the primitive way of knowledge systems. This is a model that falls short of producing adequate and useful knowledge. Yet in our contemporary, post-industrial society, we continue to cling to such ineffective models of producing knowledge.

Why do we continue with our misplaced confidence in this primitive model? Among the possible reasons, I propose the following four key factors. First, since the middle of the twentieth century, we have accumulated and have become accustomed to an enormity of data and information. We designed and constructed massive warehouses—physical and electronic—to store and to manipulate this massive collection. We currently, for instance, have information on almost all working people, including their credit history, economic and financial activities, and their health and employment experience. To the chagrin and desperation of supporters of individual liberties and of privacy advocates, almost all private and public organizations are in possession of even minute details of our lives. In short, as a society, we have too much data and information at our disposal, and we continue to collect and store an ever-growing quantity of whatever information we are allowed to gather.

Secondly, the proliferation and increasing sophistication of digital and analog computing has allowed us to develop technology for the manipulation of such data and information bases. Although quite rudimentary, this technology gives us the illusion that we are in control of the mass of information. The technology also provides us with analyses of trends and other statistical attributes of the enormity of information, so we can *locate, find,* and *extract* specific information out of the burgeoning ocean that we possess.

For example, if Jane Smith applies for employment at the marketing department of a middle-sized company, the recruiting manager can access, within minutes, Ms. Smith's life history, her credit experience, and her purchases at Bloomingdales, Walgreens pharmacy, and any encounters with law enforcement since Ms. Smith had obtained her driver's license. By her consent, the prospective employer is able to acquire information about Ms. Smith's grades in high school and college, all the medical procedures she had since childhood, and a multitude of other information. The prospective employer still does not know who Jane Smith is, but he has all the information about what Jane Smith has been doing in her entire adult life.[1]

Thirdly, the history of information technology and information systems in organizations points to the origin of this massive handling of data and information in what is known as the "back-room" of the organization. Data processing (as it was originally known) started in such organizational functions as accounting (accounts payable and receivable), payroll, and inventory control. In these functions it was sufficient to have a system that could collect and process massive amounts of renewable data, and to obtain trends, statistics, and behavior of the data over time and by individuals and groups. The knowledge embedded in these databases was of no interest to the discharge of such functions as paying creditors, managing inventories, and meeting the organization's own payroll.

Finally, we seem to hold on to the primitive model because the move from information systems to knowledge systems was in most organizations, at best, a fad, and usually just a renaming of systems of tested practices with which everyone was comfortable and familiar.[2] The change to knowledge management systems (KMSs) was a slow process, driven by consultants who were primarily experts in the traditional *information* systems. These "guides" to corporations and to government organizations had to practically build the knowledge system "as they went along," mostly by trial and error.

From Information to Knowledge Systems: The Tortuous Road

In the final analysis I can confidently state that we are now (at the dawn of the twenty-first century) only at the dawn of the knowledge society. Knowledge systems and their management are in a nascent stage. There is a persistent gap between what the experts promise that knowledge systems can deliver and what these systems truly engender for their organizations. We are all squarely in the midst of what boils down to be "glorified information systems."[3]

As early as 1971, C.W. Churchman had argued that knowledge in organizations should be more than just a "collection of information." He added that "knowledge resides in the user and not in the collection" (Churchman, 1971). However, more recently the definition of knowledge management does indeed promote the role of information, with some reference to tacit knowledge. For example, a prominent consulting group such as the Gartner Group has the following definition: "Knowledge management promotes an integrated approach to identifying, capturing, retrieving, sharing, and evaluating an enterprise's *information assets.* These information assets may include databases, documents, policies, procedures, as well as the uncaptured tacit expertise and experience stored in individuals' heads."[4]

The experts in knowledge management are inventing the transition from information to knowledge as they go along. This effort seems to be hampered by at least three categories of barriers. The first is the location of much of the volume of explicit organizational information in documents and databases. Research in managerial and organizational sciences that relies on extracting findings from these sources refers to them as "unobtrusive measures."[5] Although we usually believe that by having records, documents, and databases, we are ready to extract desired information, the reality with such unobtrusive sources is much harsher and unyielding. Records and documents present a variety of difficulties when we access them and wish to extract the information they contain. For example, people in organizations create and record information for many reasons: political necessity, organizational procedures, the need to cover one's track, to justify one's actions, and to present a complimentary image to others and to posterity. Records, such as reports, memoranda, and even analyses of events and trends, are sensitive to time and dependent on the writer's education, skills, and abilities to capture the phenomenon being recorded. Writers describe facts and events as *they* see them at that particular moment in time. They use terminologies that are specific to their profession, organizational climate, or a certain period with its fads and preferences.

In addition, we are also confronted with the biases embedded in archival information. People who create these records tend to deposit only the information that suits their needs, allays their fears, and in general, simply satisfies their obligations to the organization. They tend to ignore or leave out information that may compromise them in any way or that fails to satisfy their needs. It is therefore up to the searcher of such records not only to access these archives, but also to *interpret* what the original creator was attempting to accomplish by creating the records and, if at all possible, the story embedded in them, somewhat stripped of the biases and subjectivity.[6]

A second category of factors that generate impediments to the transition from information to knowledge is a host of *organizational* barriers.[7] They include the reluctance of organizational members to deposit their knowledge in knowledge management systems, and the less than successful track record of incentives designed to promote usage of these knowledge systems. Other barriers are the problems associated with implementation of knowledge systems. Many organizations install these systems without adequate training and the preparation of their employees for a change process such as the adoption of a knowledge management system.[8]

A third category is the set of definitions we generally use for information and for knowledge. The reader will bear with me as I again divert the narrative to a philosophical discussion of these definitions. In the general effort to install knowledge management systems in private and public organizations, there is a glaring lack of a workable definition of knowledge.[9] The recurrent theme in the definitions that do abound is the sources where information can be found, for instance, in records, documents, stories, and the memory of organizational members.

But, data and information describe events and entities in the environment. These descriptions are shared and transferred to others by means of language. This is an artifact used to transmit and transfer information. The content found in these sources and shared with others is not, and cannot be, defined as knowledge. We read such records, we count the figures provided—but we are merely defining information. This is also why we often feel overwhelmed by the information avalanche to which we are constantly exposed. The road from the information around us to knowledge is a dead-end![10]

In this book "knowledge" is the result of the clustering of sensorial inputs, not the translation or transfiguration of information. Knowledge resides in the mind of the knower who may share some of it with others. Within the framework of sharing and diffusion of whatever knowledge can be transferred, the individual transcribes concepts and ideas in the form of statements or propositions, so these can be captured by others.

For example, these may be items of information in the physical environment, such as attributes of the winds, clouds, time of year, color of the sky, humidity, and animal behavior. A person may consider all these elements and share the following notion: "IF all these items are present, THEN there is a tornado brewing in the West."

Even when such knowledge is recorded, there is little else that can explain how the individual arrived at such knowledge. Sometimes it requires a lifetime of experience to explain the clustering of sensorial inputs which end up as knowledge.

In summary, there is an undercurrent of the design of knowledge systems which are essentially collections of information systems and offer little that can be characterized as workable and usable knowledge. We also encounter a host of organizational barriers when we attempt to implement and adopt such knowledge systems (Lehaney, Clarke, Coakes, & Jack, 2004).

The Good, the Bad, and the Promise of Data Mining and Knowledge Discovery

Data mining is usually defined as the collection of several techniques utilized to analyze very large bodies of data with the objective of providing structures, patterns, and models in the data. Data mining (DM) usually relies on such techniques as mathematical computation and machine learning, and uses principles of association, segmentation, and classification of the data. The term data mining has been favored by information systems professionals. But in the early 1990s a new term was coined by analysts who wished to show a connection between finding patterns in large volumes of data and the knowledge that may be embedded in them. The term was *knowledge discovery in databases* (KDD) (Fayyad, Piatetsky-Shapiro, & Smyth, 1996). Fayyad and his colleagues at Microsoft, GTE Laboratories, and the Jet Propulsion Laboratory suggested that data mining is just a step in the KDD process, and that the other components of KDD include data cleaning, assessing prior knowledge, and the interpretation of the results from the mining effort. KDD is therefore defined as the *process* of knowledge extraction and interpretation from the databases being mined.

However, are these patterns that the KDD process identifies indeed *knowledge*? The analysts proposed a limited definition to what they consider to be knowledge: "We can consider a *pattern* to be knowledge if it exceeds some interestingness threshold which is by no means an attempt to define knowledge in the philosophical or even the popular view…Knowledge in this definition is purely user oriented and domain specific and is determined by whatever functions and thresholds the user chooses" (Fayyard et al., 1996, p. 41).

Herein lies the problem. Data mining works by applying models to databases in order to extract the data that fits the model. For example, a pharmaceutical company wishes to tailor sales of a new drug to clinics and physicians who treat patients with the illnesses targeted by the new drug. The company will model the information it desires in this case, such as the particulars of the targeted clinics and physicians, characteristics of their patients, and levels of prescribing activity by these physicians (in general, and for drugs sold by the focal company). Armed with the model, the company will mine its own databases of past sales with other drugs and databases from other sources, such as regulators, insurers, and the government.

From a business perspective, thanks to data mining we now have specific information on where to target sales effort and how best to utilize the sales force. This is the good side of data mining. To obtain such results we needed to have a prior model of what we want to extract and in what format, and some preparatory work to "clean" the data being mined. We may therefore argue that the patterns we discovered are those we "predicted" or hoped and planned to find. Otherwise we would be operating in the dark, and any patterns we uncovered would have been random and of no or little use.

But, is this knowledge, and can we now incorporate such patterns and their interpretation into a knowledge system? Even users of data mining and KDD admit that their definition of knowledge is restricted to how the user perceives the usefulness of the analysis. Again, we are all operating in a universe of data manipulation without an acceptable definition of what constitutes knowledge.

The promise of data mining is in the useful manipulation of masses of data to glean some structure, trends, and patterns that can be used to improve organizational operations, processes, and performance. The more such promises are kept, the more we find ourselves immersed in a race to gather more data, improve the mining of these data, and search the databases for other patterns. There is not much in the experience we had with data mining to categorically show that there has been a transition to knowledge discovery.

Perhaps this is simply an outcome of the lack of a unifying definition of knowledge, or because of too many definitions of knowledge tailored to specific needs, aims, and methodologies. The experience we currently possess in data mining is concentrated on information processing, not on knowledge management.

It is however abundantly clear to me that data mining is an essential tool for large organizations to explore, to access, and to make sense of their databases. This capability is a very valuable support for making decisions and for strategic survival. But, it is not knowledge.

Toward a Neuronal Model

Instead of the primitive Model A, I am suggesting here a different Model B, or the *neuronal* model of knowledge. In this model the emphasis is on the principles of structure and progress that I outlined in this book.

In Model B the structure of knowledge is not by semantics or semiotics, but by clustering of sensorial inputs. Consider the vast array of information on the Internet. Presently search engines scan the Internet by the principle of finding key terms and the presentation of relevant Web sites, ranked by some criteria of importance to the researcher. In Model B we search for *knowledge* embedded in the database. This means that we *cluster* terms and concepts as they exist in the data warehouse. But it also means that we need to *organize* the inputs we collect in this system in such a way as to allow for subsequent clustering. In other words, we must enter inputs into the system in a mode similar to how our brain collects sensorial inputs and clusters them before entering this outcome into the memory banks of our mind.

Imagine a search engine that locates such terms or concepts as David Hume, Immanuel Kant, and theory of knowledge. The engine would cluster these terms and will be able to answer questions such as: (1) What are the key arguments of the theory of knowledge of these scholars? (2) How different is Hume's theory from Kant's? (3) And are there any similarities between them, and if so, what are they?

Evidently we would need to cluster more than the three terms listed above. We would need to cluster terms or concepts such as "arguments," "similarities," and "differences." But here enters the second principle of Model B: *progress* of knowledge. This principle is based on the critical role that the *backbone* of the knowledge base plays in the model of progress, and the *growth* pattern of the knowledge base, particularly at the margin.

All knowledge is relevant and can be called upon at any time to be incorporated in a clustering effort. The knowledge base should include such terms or concepts as "similarities," "differences," "a priori," "a posteriori," and other attributes and constructs of the concept of theory. So, when we now add to the backbone the contributions of a David Hume or an Immanuel Kant, we are adding to the margin and drawing from the body of knowledge we built over time.[11]

Model B is designed to emulate the human brain in the modes of creating knowledge and accessing it. The principles of clustering and progress are meant to work with clustered conceptual constructs rather than with raw terms. This mode of knowledge management may require new designs of computers and search mechanisms that operate like our brains. Multiple inputs which undergo clustering are the norm, not the exception. Just as the brain clusters multiple sensorial inputs, knowledge systems will cluster multiple inputs, some in the form of lower-level constructs, others in the form of higher-level concepts. The intelligent searcher should be able to, at the very least, ask the system questions that elicit *knowledge,* such as why and how, and compare and draw conclusions.

The Post-Lyotard Era of Knowledge

Model B can also be named the post-Lyotard Model, after Jean-Francois Lyotard (1924-1998). In his description of the post-modern condition of contemporary society, Lyotard had argued that "narrative knowledge" has been replaced by "scientific knowledge," which is "self-legitimatizing," based on criteria of efficiency and utility in its exchange value.

To build upon Lyotard's ideas, contemporary society with its information technology and computerization has changed the way we collect, store, and access knowledge. We do so (as I elaborated above) on the basis of semantics and symbols. We essentially abandoned the basic mode in which people exchange knowledge, which is by storytelling, free narratives, concepts, and clustering of these inputs.

The advent of technologies and the scientific method have transformed human exchange into a terribly constrained process of efficient transfer of dedicated specifics and well-defined words and symbols. Modern philosophers of knowledge have established the belief that one cannot "know" the true world or true reality; therefore knowledge can only be explored in the realm of manipulation and exchange of language.[12]

With the abandonment of storytelling, allegories, and the free exchange of broad narratives, we have lost the edge in human exchange. The mode of "general narratives" allowed people to "pick and choose" from the narrative those aspects of it (terms, concepts) they wished to cluster and to further act on them—similar to how we scan our environment with our senses and pick-and-choose sensorial inputs for clustering and the generation of knowledge.

In the context of organizational life, we have abandoned the notion of exchange of knowledge via storytelling and free narratives. Managers are required to share what they know in the form of short, standardized reports. In one company for which I had consulted, the president instructed the director of marketing to "talk about your experiences with the marketing of these products—you have six minutes." The proverbial "water cooler" around which organizational members chat and exchange stories is frowned upon by many organizations. The advent of electronic mail has significantly curtailed direct, interpersonal conversations. In this regard, a U.S. Appeals Court ruled in February 2007 that an anti-fraternizing policy of a security services company violated the rights of employees to socialize with co-workers as long as the topics of these interfaces are work related. As companies become more formal and their work processes more standardized, there is a tendency to curtail informal relations and unmonitored and nonrecordable interactions among their employees. By doing so, these companies also inhibit the exchange of knowledge among their employees.

Like the physical universe, imagine a *universe of knowledge,* where there are principles of attraction and the formation—by clustering—of "galaxies" of knowledge. There are congregations of knowledge of similar nature, correspondence of attributes, and other such reasons for association.

What is the effective mode of scanning this universe of knowledge and making sense of it—by understanding it and extracting those components in the universe that can be of use to us? Human interaction started with the ability to narrate experiences and perspectives on the world that surrounds us. Linguists have taken the elements of language and distilled them to words, their specific meaning, and their value—all of this outside the context of the story, the narrative, and the body of knowledge that supports the narrative.

Why do advanced biological creatures go to the extreme effort of maintaining a memory and even "shutting down" their alertness with the activity we call "sleep"? Pause for a moment and assess the peril associated with sleep.

Higher-order animals and humans leave themselves vulnerable to predators, at great peril and for a long period of time. We sleep not because of the darkness of night, as some animals are nocturnal and sleep during the day. We sleep to allow our brain to sort through knowledge acquired and accumulated in the previous period. The brain "cleans up" the knowledge base and decides on which knowledge to store in memory. Throughout this process our brain surveys such knowledge, resulting in dreams.[13]

This is similar to my home office. It is cluttered with books, computer disks, and documents. Periodically, at the instigation of lack of space for additional material and my wife's protestation, I clean-up the office. While doing so I review documents before condemning them to the shredder or to a resting place in the basement of the house. Reading through these documents evokes memories, helps me to clarify ideas, and brings joy as well as pain—just as dreams would.

We must consider the entire catalog of knowledge in the design of an effective mode to scan and to make sense of the universe of knowledge. The scholars who advance semantics, linguistics, and semiotics as means to acquire knowledge are missing the point about the importance of the supporting body of knowledge we already possess.[14]

From Individuals to Organizations: Language, Processes, and Performance

In the contemporary, post-industrial society, there has occurred a shift from individuals exchanging their knowledge to exchange and transfer of knowledge within and through organizations and their processes, procedures, and structures. This is a major shift in the way modern people interact. Instead of individuals interfacing directly with other individuals, much of the exchange and processing of knowledge occurs within and through organizations. This is true for the workplace, where within and between organizations the knowledge exchange and transfer has become highly institutionalized, well structured, and devoid of the individual aspects of personal interaction.

A key factor in this phenomenon is the belief in work organizations that formal procedures and institutional mechanisms are adequate solutions for almost every problem including exchange and transfer of knowledge. But these mechanisms have repeatedly failed to extract and to utilize tacit knowledge of organizational members.

This is why although frowned upon by organizations, workers are still enchanted by the concept of gathering around the proverbial "water cooler" where they exchange stories, latest gossip, and other tidbits of knowledge about life in their organization. It is the so-called *informal organization* identified in the 1930s by Elton Mayo and other researchers at the Western Electric Hawthorne Plant near Chicago. The existence of the organizational "grapevine" helps people to overcome their reluctance to confide in formal structures and to interact more freely with other workers without fear or constraints (e.g., Cobb, 1977).

In many respects these organizational mechanisms have taken the place of language as the mode of knowledge exchange. Language, by means of stories, narratives, and a fluid interaction among individuals, is a powerful mode of exchanging tacit knowledge. Woven into these narratives are feelings, special thoughts, emotions, and other such expressions of what we know and how much importance we attach to this knowledge. Interpersonal exchanges involve *trust* and *sharing.*

Conversely, formal exchanges are impersonal and are focused on the proliferation of facts and data, rather than knowledge. Workers are concerned about the negative reaction from managers and the depositing in long-term institutional memory of whatever they disclose.

Organizations believe that they can improve their performance by transitioning from the level of interpersonal exchange to the more complex institutional level of the organization's knowledge management system. This is partially true, and it is why knowledge experts overpromise and knowledge systems fail to deliver. The standard knowledge management system is impotent, inconsequential, and devoid of the richness embedded in the experience and lifelong accumulated knowledge ensconced in the minds of organizational members.[15]

Yet, knowledge management systems are designed to impact organizational functions, activities, and processes. This indirectly improves organizational performance and strategic

survival. Although, in my view, the field of knowledge management is still in a stage of disarray, some benefits are even currently accrued to those organizations that took the initiative and installed knowledge systems. In particular, the benefits begin to show when these systems (viewed as corporate knowledge *assets*) leverage the knowledge they already possess in their systems and processes, identify such knowledge, and manage its usage in an *integrated* fashion. These actions alone are enough to generate some measurable benefits from the existing pool of knowledge. But it is a far cry from what is being promised and what can be accomplished. Model B is a conceptual starting point for a world where such promises can be fulfilled.

Towards a Paradigm Shift in Knowledge Management Systems

Knowledge management systems are at the present time artificial structures designed to receive, store, and exchange knowledge. Since the publication of the work by Ikuiro Nonaka and his Japanese colleagues in 1995, the approach most commonly used with regard to KMSs has been to improve the movement from "tacit" to "explicit" knowledge. In practical terms this meant the application of individual and organizational incentives to make people delve into their pool of tacit or embedded knowledge and then to externalize this knowledge. Nonaka and his followers believed that the Japanese experience is transferable to other cultures and that organizations in any environment would be able to create conditions that would be conducive to their members to share and diffuse what they "tacitly" know.

With over a decade of experience with this approach, the results leave much to be desired. In my opinion this approach has failed, and its fulfilled promise has contributed to a general sense in work organizations that KMSs are either a passing fad or another instrument of executive control that is destined, in time, to self-destruct.

A key reason for the lack of success of KMSs outside the Japanese experience was not only the underdeveloped climate of teamwork in the Western work culture, but also because the approach to entice the movement from tacit to explicit knowledge has not worked even in Japanese companies, which Nonaka and his colleagues had studied. Perhaps what Nonaka had observed was not tacit knowledge being transformed into explicit knowledge, then shared and diffused to others, but explicit knowledge regurgitated and reinterpreted by Japanese managers.

As a human phenomenon, only in situations of intensive interpersonal interactions (such as mentoring) do individuals share some of the knowledge embedded in them—"tacit" knowledge. To evoke the knowledge embedded in the mind, there must be a personal involvement by the knower, combined with the opportunity to be engaged with another person and to exchange—interactively—some of the knowledge one possesses. This is true for all types of knowledge in the Nonaka and Takeuchi classification. When, then, would a person in the context of the work organization confide such embedded knowledge to a cold, impersonal, and institutional system: the KMS? In addition to the nearly impossible task of evoking tacit knowledge and transforming it into explicit knowledge, there is also a host of other personal and organizational factors that inhibit the interaction with a KMS and which I listed previously in this chapter.

So, if this approach of transforming tacit to explicit knowledge is ineffective, what can be done to improve the working of organizational knowledge systems? The approach I suggest here is a shift in the paradigm, or the way of thinking, which until now has dominated the KM field.

By accepting the *neuronal model* and the reality that all knowledge is tacit and only a very small portion of it can be shared and diffused, the emerging and different approach focuses on the KMS as an *organizational* rather than a knowledge system. Instead of searching for methods that will elicit tacit knowledge, the emphasis shifts to making the KMS more receptive to interaction with organizational members, and more efficient in overcoming the barriers to the depositing and extraction of knowledge from such systems. The problem shifts from how to make organizational members share and exchange their tacit knowledge to how to encourage them and facilitate the sharing of knowledge in general, in this case, overwhelmingly explicit knowledge.

For example, a large manufacturing company had undergone a major reorganization of its engineering department. The company had in operation a rudimentary KMS. In order to facilitate the transfer of knowledge from the current managerial staff to recent hires, the company provided incentives to its managers to enhance tapping their "experience" or tacit knowledge.

Several of these managers targeted for knowledge transfer had also been targeted for separation or termination of employment, once the reorganization reached completion. Very little, if any, knowledge had been transferred from the existing cadre of managers into the KMS, or directly to the new cadre of managers.

The company then changed its approach. It provided incentives for existing managers to form joint working groups with new hires, and to discuss company issues and challenges within these committees, thus bypassing the KMS and refraining from asking managers to share and to diffuse their personal knowledge. This approach proved to be much more successful because it facilitated a markedly increased level of interaction and a more intensive exchange of explicit knowledge, peppered with some tacit knowledge. As an illustration, in one such joint team, a long-time manager commented: "This will not work. We tried it over a decade ago and we failed. I think I know why. Let me explain...."

In a way, the failure of the current approach is similar to the criticism of the search for unity in the physical sciences (Woit, 2006). The failure is due to a basic flaw in the rationale for the approach and the effect of several other factors which impede its success and are inherent in its structure.

The Knowledge Ratio and Concluding Thoughts

The neuronal Model B is designed with the objective of improving the working of knowledge systems so they can better emulate the human brain. The idea is that if we are able to build systems that emulate the way the brain structures knowledge, we will open immense possibilities to artificially create and manipulate knowledge—beyond our existing databases.

Consider a ratio of knowledge that measures the levels of tacit and explicit knowledge embedded in any knowledge system.

$$\text{Ratio of Knowledge (ROK)} = \frac{\text{Explicit Knowledge}}{\text{Tacit Knowledge}}$$

The notion of tacit knowledge is also *declarative* knowledge, or the volume of knowledge and experience we have accumulated in our lifetime. The ratio of knowledge provides an approximation of the degree to which the knowledge of which we are aware and which has been publicly exposed is a part of tacit or declarative knowledge of an individual knower or a knowledge system in an organization.

Another definition of the ratio would be: "How much knowledge do we really have in databases, processes, procedures, and other such components of the organization, and how much knowledge is still beyond our reach, embedded in the minds of our people?"[16]

In managerial terms the ratio of knowledge may be a tool to evaluate the power of a knowledge system in the organization. The problem, of course, is how to determine what is tacit knowledge. There are ways to make an approximate assessment by evoking, for instance, the knowledge embedded in organizational members through *debriefing*. When I conducted interviews in large organizations, I discovered that managers possess much more knowledge about the topic we were discussing than they were willing to share or to enter in organizational records. I also discovered that managers are generally amenable to discuss and to share some, but not much, of their experiences.[17]

I would roughly guess that the ratio of knowledge in most large organizations is 1/10, so that about 10% of the total corporate knowledge is explicit in the records and archives of the organization. Yet, even though knowledge management is a nascent discipline, there is hope for KM and KMSs. Organizations are populated by clever and talented people who routinely use their tacit knowledge to solve problems and to "get things done." In smaller companies there is a more pronounced rate of sharing knowledge, perhaps because of better familiarity among members, lack of formal constraints, and less fear of negative feedback from internal competitors.

So, the key to successful knowledge management systems is to design them with people in mind, and to make them *enabling* instruments that facilitate human interface, hence to be able to increase the sharing of any knowledge, starting, of course, with explicit knowledge.

There always was knowledge, since the beginning of civilized living and the sharing of tasks in societies. Are we discovering the obvious, now that we are suddenly and dramatically venturing into the indispensability of knowledge systems? Not so! We are dragged into the inexorable need to learn more about knowledge and to harness it—because of the pressures of modernity, competition, and the enormous achievements in science and technology.

Knowledge is no longer simply a tool in the conduct of human affairs. It has now become the engine that drives economic and social activities and human interaction and performance.

The "Dual Phenomena" of Knowledge

As human beings we create knowledge from sensorial inputs, we store this knowledge in our mind, and we are only able to transfer and share a small portion of what we "really" know. The ability of our mind to create and use knowledge is strictly an individualized mode of survival dictated by evolutionary exigencies and allowances. The social mode of human existence and operation within groups and even larger organizations is a relatively new development and in many ways an exception. In lower animals, such as insects, the social structure relies on the sharing of *all* knowledge that each member has and brings to the collective. Although humans have been able to develop such institutionalized communities, they nevertheless still find it very difficult to communicate and to share their knowledge. To do so requires much effort and skill.

The duality of having knowledge for individual purposes, and knowledge for existing in social institutions and contributing to them, constitutes a major human achievement and at the same time also a major challenge. How well we can confront and overcome this challenge will help to determine how well we make our knowledge society and knowledge economy bring us all the good fortune we believe it encapsulates.

The purpose of this book is to bring into the popular domain the notions and applications of knowledge and knowledge systems. While attempting to do so, I also interjected in the narrative my personal views of how knowledge is structured and how it progresses. I believe that the field of knowledge management is progressing, albeit cautiously, and that we will continue to improve the effectiveness of organizational knowledge systems. All is contingent upon bringing organizational principles into alignment with how people think, and how they create and exchange what they know and what they have experienced throughout their lives.[18]

References

Churchman, C. (1971). *The design of inquiring systems.* New York: Basic Books.

Cobb, A. (1977). *Informal influence in the formal organization.* Berkeley, CA: University of California Press.

Fayyad, U., Piatetsky-Shapiro, G., & Smyth, P. (1996). From data mining to knowledge discovery in databases. *AI Magazine, 17*(3), 37-54.

Lehaney, B., Clarke, S., Coakes, E., & Jack, G. (2004). *Beyond knowledge management.* Hershey, PA: Idea Group.

Rubenstein, A.H., & Geisler, E. (2003). *Installing and managing workable knowledge management systems.* Westport, CT: Praeger.

Webb, E., Campbell, D., Schwartz, R., & Sechrest, L. (1966). *Unobtrusive measures: Nonreactive research in the social sciences.* Chicago: Rand McNally College.

Endnotes

[1]   This includes air travel and the technology to scan fingerprints, one's retina, and other biomedicine attributes that help authorities identify and single out individuals for the purpose of air travel security and soon for other purposes.

[2]   There is growing literature on the shift from *information* to *knowledge* systems and management. The reader will be interested in: Maier, R. (2004). *Knowledge management systems: Information and communication technologies for knowledge management* (2nd ed.). New York: Springer-Verlag; and Torsun, I. (1995). *Foundation of intelligent knowledge-based systems.* New York: Academic Press. Also, Rothberg & Ericson. (2004). *From knowledge to intelligence.* Oxford, UK, Butterworth-Heinemann.

[3]   The reader may wish to consult the following references in the knowledge management literature: Prusak, L. (2001). Where did knowledge management come from. *IBM Systems Journal, 4*(2), 1002-1007; White, D. (Ed.). (2002). *Knowledge mapping and management.* Hershey, PA:

Idea Group; and Malhortra, Y. (2001). Expert systems for knowledge management: Crossing the chasm between information processing and sense making. *Expert Systems with Applications, 20*(1), 7-16.

[4]   The emphasis is mine. The definition can be found in various reports by Gartner on knowledge management.

[5]   The reader may wish to consult the seminal book by Webb, Campbell, Schwartz, and Sechrest (1966).

[6]   Webb et al. (1966) succinctly summarized these barriers: "We should recognize that using the archival records frequently means substituting someone else's selective filter for your own" (p. 111). But they also cite a Chinese proverb to show their belief that regardless of the difficulties, archival information is invaluable: "The palest ink is clearer than the best memory."

[7]   The reader will find an extensive list of these barriers in Rubenstein and Geisler (2003).

[8]   See, for example, Mahlhotra, Y. (2004). Why knowledge management systems fail. Enablers and constraints of knowledge management in human enterprises. In Koenig, M., & Srikantaiah, K. (Eds.), *Knowledge management lessons learned: What works and what doesn't* (pp. 87-112). Silver Spring, MD: Information Today. Also Ward, J., & Peppard, J. (2002). *Strategic planning for information systems.* New York: John Wiley & Sons.

[9]   In our book (Rubenstein and Geisler, 2003) we defined knowledge in terms of its components as "nuggets."

[10]   The issue seems to be the need to define *knowledge* beyond "keywords" and "terminology" offered by language and semantics. In the philosophical literature there is an extensive discussion of this topic. Saul Kripke, for example, had argued that mental facts are not identical with physical fact. He opposed the *descriptivist* approach to reference (how language describes objects in the physical world). He suggested instead that we name an object in the physical world because of a *causal connection* we have with this object. He said that there are *rigid designator*s and will refer to the object in all the worlds where the object can exist. This theory differs radically from the descriptivists (such as Bertrand Russell) who argued that we name an object because the term is part of a description of the object. See: Kripke, S. (1980). *Naming and necessity.* Cambridge, MA: Harvard University Press. In my forthcoming book,

The Treatise of Knowledge, I reject these arguments in naming objects. I suggest instead that linguists and semanticists approached knowledge in a way that started with higher-order constructs and propositions, thus ignoring the building-blocks of knowledge and their relation to the physical world.

[11]   Readers may call this a "learning process," and indeed it may well be that such a system learns by adding knowledge to an ever-growing body of knowledge. Regardless of the notion of learning, the neuronal model is designed to emulate the generation of knowledge in the brain.

[12]   For reasons of brevity I over-simplify the key arguments of such philosophical perspectives as the logical positivists: Wittgenstein, Carnap, Schlick, and more recently Kripke. See, for example, Ayer, A. (1978). *Logical positivism* (reprint ed.). Westport, CT: Greenwood.

[13]   Recent findings in brain research have suggested that rapid mutation that led to the expansion of the human brain has helped the survival of humans in the evolutionary struggle. Humans had separated themselves from primates due to an accelerated evolution (via mutations) of the subset of genes that are responsible for the development of the brain and the nervous system. Thus, the human brain became larger and more capable of processing and storing larger stocks of knowledge—without having to wait for the regular selective process of evolution, which would require millions of years to take place. We humans went to all this trouble to create a larger brain for the more effective processing of a large body of knowledge—we should not therefore ignore the role that such a body plays in processing marginal new knowledge. For the study on the mutation in 17 brain-developing genes, see: Dorus, S. et al. (2004). Accelerated evolution of nervous systems genes in the origin of *Homo sapiens. Cell, 119*(December 29), 1027-1040.

[14]   For example, the reader may wish to consider the case of digitized fingerprints. In January 2005 the *Chicago Tribune* reported that the database of fingerprints of the Federal Bureau of Investigations (FBI) has been digitized. This means that any set of individual fingerprints can be compared with the database and a match can be established within minutes, rather than weeks. The unfortunate side effect is that the computerized technology has imperfect resolution of the picture, and thus may lead to false accusation of innocent people and to overlooking the guilty. This outcome is possible when fingerprints are the *only* item considered in the determination of guilt or innocence. If a more

varied catalog of knowledge about the individual suspected of a crime is also considered, the negative effects of one imperfect item will be negligible.

[15]   The reader may wish to consult: Firestone, J. (2002). *Enterprise information portals and knowledge management.* Burlington, MA: Butterworth-Heinemann; McElroy, M. (2002). *New knowledge management: Complexity, learning, and sustainable innovation.* Burlington, MA: Butterworth-Heinemann; and Barquin, R., Bennet, A., & Remez, S. (2001). *Knowledge management: The catalyst for electronic government.* Vienna, VA: Management Concepts.

[16]   Knowledge ratios have been used by government agencies and corporations to denote ratios of knowledge workers and knowledge bases needed to run the organization. For example, the firm's knowledge ratio =

$$\frac{\text{number of knowledge workers employed by the company}}{\text{total number of employees}}$$

A similar ratio for an industry or sector would be:

$$\frac{\text{number of knowledge workers in the industry}}{\text{total number of employees}}$$

For instance, the nuclear industry would require a higher knowledge ratio to run the nuclear reactors than the fast food or the apparel industries. However, the ratio of knowledge proposed here refers to the *knowledge* itself, not to the knowledge workers.

[17]   The reader will find a more complete description of the power of debriefing in my pages: Geisler, E. (1999). Harnessing the value of experience in the knowledge-driven firm. *Business Horizons,* (May-June), 18-26.

[18]   The reader may wish to consult: Shannon, M. (2003). Knowledge conversion is the key to success. *The Information Management Journal,* (November-December), 52-57. The author concluded that "those companies that facilitated the converting of information to knowledge did better than those that only provided hardware and information conduit services" (p. 57). Also: Geuna, A. (1999). *The economics of knowledge production.* Cheltenham, UK: Edward Elgar.

Chapter XVI

SCIO, ERGO, OMNIS:
The Knowledge Perspective of Everything

On the Knowledge Perspective

In this book I brought forth notions and views about the structure and progress of knowledge that are different, perhaps even revolutionary. Such was not my original intention, just an unintended consequence or a resounding side effect of my framework of what constitutes knowledge and how it grows.

In a wonderful little book, Aleksander R. Luria (1902-1977) studied the peculiarities of a person with vast memory, the mind of a mnemonist.[1] His patient was able to memorize details, but was unable to understand metaphors and poetry, and he lacked conceptual thinking. The patient lived in a world of particulars, unable to transform them into generalized notions. Although he was able to blend his sensations, he nevertheless had great difficulty in transitioning these sensations into abstract thinking. Luria's mnemonist had to visualize everything in order to comprehend abstract concepts. This resulted in a constant struggle to, as Luria said, "arrive at meaning through these visual forms" (p. 135).

This is a case of an extraordinary person with the capacity to capture data and information but not knowledge. The inability to cluster sensorial inputs and images into meaningful notions has caused this person much pain and difficulties throughout his life. In essence, it is not the knowledge we possess that is so crucial to human functioning and human survival, but the ability to *generate* knowledge, by clustering sensorial inputs and by dependence on an existing body of accumulated knowledge, previously generated and clustered.

Consider another case of people who suffer from prosopagnosia ("face blindness"). This hereditary disorder affects, to some degree, about 12% of Americans. Patients who suffer from it are unable to differentiate and to recognize faces, even those of relatives or even their own face in more severe cases (Kennerknecht et al., 2006). Although their vision may be excellent and they are also able to discern voices, clothing, and hair color, they are unable to cluster the sensorial inputs needed to create the knowledge of face recognition.

Regardless of the information available to the people who suffer from this disorder or the richness of sensorial inputs they receive, the cognitive process of clustering these inputs into knowledge is deficient. These patients cannot describe a face they have just seen, or recognize the face (acquire knowledge about it) from communication devices such as pictures, narratives, videos, or documents. The knowledge they possess (or in this case they lack) about faces is embedded in their mind and is a *personal* experience or absence of such experience. About six million Americans suffer from this disorder in one form or another.

The enormity of the role of knowledge in human existence makes it the perfect candidate for being the link we create with the world around us. The prism or lens of knowledge is the perspective with which we view everything. It is what and how we know that determines: (1) that we exist, (2) who we are, (3) what we can do, (4) what we mean to ourselves, (5) what we mean to others, and (6) what life is all about. Knowledge is not only a crucial element in our link to the world, it is the only prism to give meaning to our world. Without knowledge and the ability to generate it, we would merely be a poor imitation of a mnemonist—with some memory and no means of making sense of what we observe and experience.[2]

In this chapter I have gathered some possible perspectives of the human condition, seen through the lens of knowledge. Like Alice in Wonderland peering through the looking glass, we can view reality through the prism of knowledge. The starting point is the proposition *Scio, ergo sum* (I know, therefore I am). From this I extend the proposition to *Scio, ergo vita* (I know, therefore life) and *Scio, ergo omnis* (I know, therefore everything).

Knowledge in these perspectives is not an explanatory variable. It does not explain life, existence, or the self. It *defines,* in very selective and unique terms, who we are, what we mean, and what are life and existence.[3]

SCIO, ERGO SUM: Knowledge and Existence

In 1637, Rene Descartes (1596-1650) published his philosophical treatise, *The Discourse on Method.* He compared human knowledge to a house. In order to build a better and lasting structure, he argued that one must tear down the existing structure. Descartes classified his beliefs into three groups: beliefs that originate from the senses, beliefs from imagination, and beliefs from understanding. He doubted the belief framework that existed at that time and argued that *all* these beliefs needed to be re-examined. Such inclusive doubt led him to proclaim that although he may question all knowledge, the fact that *he* is thinking and doubting is a true existence. Descartes coined the phrase *cogito, ergo sum* (I think, therefore I am).

Almost two millennia before Descartes, the Greek scholar Socrates (469-399 BC) supposedly declared *scio me nihil scire* (I know that I know nothing—meaning certain knowledge cannot be attained).[4] Socrates emphasized the necessity and the challenge in searching for knowledge, albeit at times it can be a frustrating endeavor.

So, building upon the ideas of these scholars, I contend in this chapter that if knowledge is the perspective of everything, then *scio, ergo sum* (I know, therefore I am). This statement may have been previously written by Jean-Paul Sartre (1905-1980). This is possible, since bringing together the notions of knowledge and being transports us to the realm of the philosophical school of *existentialism.*[5]

What are the tenets of existentialist philosophy? In Sartre's words: "Existence is prior to essence…Existentialist despair and anguish is the acknowledgment that man is condemned to freedom…Man is nothing else but that which he makes of himself…Consciousness is 'nothing' but the self is always on the way to being something."[6]

On knowledge, Sartre argued that "man is defined by what he can know" and "knowledge is neither a relation, a quality, nor an activity; it is the essence of the for-itself." Therefore, knowledge is viewed as a tool (albeit imperfect) that permits the individual to make choices. If, as existentialism decrees, the individual is left in life to his own devices, alone in a world not of his making, with no apparent reason or purpose for being, then he needs knowledge of the world and of himself to freely make choices. No wonder this individual will be defined by the knowledge he can garner. Nothing else can be used to define his existence, except existence itself.

For the existentialists, existence comes before knowledge. Hence, our experience as an existing someone for ourselves and in the experience of others precedes the knowledge we accumulate and which will then define us. I contend that knowledge is beyond experience: *scientia experior practer est.* Knowledge does not *affirm* existence; it is merely a lens through which we can frame elements of our human existence within a unifying structure that can be applied across such elements. This lens allows us to frame higher-level constructs such as *God, truth,* and *consciousness,* as well as serve as the mechanism by which we cluster the basic sensorial inputs.

"I know, therefore I am" is a declaration of both structure and progress. I am because I cluster my sensorial inputs into generalized forms of understanding. I am also because I accumulate a body of such knowledge and have it at my disposal so that I can exist, survive, and succeed in whatever I choose to do. In this sense individuals are not alone in a threatening world, rather they are well equipped with the ability to create and to store knowledge. Knowledge is both a *principle* of existence and a *tool* in successful manipulation of human existence. This unity of purposes makes knowledge a defining aspect of the self, as well as everything else in his world.[7]

Two complementary questions come to mind: do we exist beyond (or independent of) knowledge? And, do we ever *really* know?[8] I argue that knowledge is generated subjectively, within the individual's mind. Therefore one has to exist before any knowledge can be created.[9] Existence and knowledge are intertwined, so that a temporal as well as a conceptual distinction is minimal. Once shared and diffused outside the individual who generated it, knowledge is transformed to an existence *beyond* human existence, because it can outlive the human temporal existence.[10]

Do we really know ourselves and our world? In my view of knowledge, it does not matter. Imagine the challenge of space travel. We can send a manned expedition beyond our solar system, and we can do so within a planned timeframe and a given budget. We can also try extremely hard and accomplish this task in half the time and half the budget we allocated to the project. Besides the elation of accomplishing this task earlier and cheaper, the task is accomplished by whatever means of time and cost. That is, we set out to go beyond the solar system and we went there.

Besides "perfect knowledge," whatever knowledge we have of ourselves and of our world is sufficient, in itself, to define our existence, regardless of whether we "truly" know the "real" world.[11]

Knowledge and the Measure of Life: SCIO, ERGO VITA

I know, therefore life. As biological creatures we begin our existence in the progenitors' genetic makeup. As individual human beings, life begins with knowledge—defined as the clustering of sensorial inputs. This is not to say that we form a consciousness, but simply that we begin to cluster sensorial inputs and generate knowledge. Such effort requires a brain mature enough to be able to cluster multiple sensorial inputs. Life, therefore, is defined and begins with knowledge.

Without the ability to cluster knowledge, we would be nothing more than a collection of molecules, organs, and a sanctuary for bacteria and other such organisms that inhabit our physical bodies. Even those human beings who, for reasons of disease, birth defects, or accident, are prevented from generating higher-order notions of knowledge possess the ability to cluster sensorial inputs to the first level of KANEs.

The measure of human life is knowledge, defined as the actualized ability to cluster sensorial inputs. There are obviously two key issues to be addressed. The first is the means we have of measuring this actualized ability to cluster. Currently we can scan the human brain to determine by imaging differentials the locations in the brain of certain cognitive and emotional activities. We are yet unable to empirically determine how clustering occurs and to what extent it has reached the construction of higher-order abstractions.

If we determine that knowledge is the measure of life, then life begins with the initial clustering of knowledge—whether we can measure it with relative precision or not. This is the second issue we need to address: when life begins and the human condition of vegetative state—the state of not having consciousness of the surroundings, what current medicine defines as "brain dead."

In the dispute whether life begins at conception or at birth, the notion of "*scio, ergo vita*" occupies the middle of the road. Until the brain is developed in the fetus to the point of being able to actually cluster sensorial inputs, the fetus is a living organism but not a "knowing organism," which is the definition of life as proposed in this chapter. This definition does not imply any measure of utility. It is anchored in the notion that *human* existence is a continuous and incessant pursuit of knowledge. This is the criterion that separates humans from other living organisms, although we share much of our genetic makeup with other fauna who inhabit our planet. As recent research has shown, we share genetic background with such diverse living organisms as fruit flies and primates. Unlike the existentialists, I reject the proposition that humans are lost in a hostile world not of their making and that there is no purpose in human life. The purpose is to pursue knowledge, both for the pleasure of the pursuit itself, and for the role that knowledge plays in our lives.

The Biblical allegory of the expulsion from the Garden of Eden is quite revealing. Humans now had to fend for themselves, and the tenets of biological evolution complement this Biblical story because in order to survive and to beat the odds of the evolutionary struggle, humans *must* pursue, acquire, and utilize knowledge about themselves and their world.[12] If, by doing so, we find this activity pleasurable, this is clearly a wonderful bonus.

The ability to produce and to manipulate knowledge *is* the rationale for human existence, hence it is *the* measure of life. It is not the connotation of existence itself. It is with and through knowledge that we exist—as human beings, distinguished from other species. Whether we arrived at this stage by design or by mutations in the development of our body and of our brain is only relevant insofar as the appearance of a larger brain compensated for other mutations that would have hindered our survival as a species.[13]

We are what we know, but we do not know what to be. This is what I would call "knowledge alienation." In a process similar to individual alienation in society, the pursuit of knowledge alienates the knower from those who have not achieved his level of knowledge accumulation. The power of rising in the chain of gaining knowledge is embedded into the evolutionary trajectory of humans.

In the final analysis, human beings are defined by what they know. Humans and their minds are merely plants that manufacture knowledge and carry inventories of previously produced knowledge. There is no inherent mission or objective guiding human existence but the pursuit of knowledge as the crucial instrument in the battle for survival of the human species, and the individual.[14]

Knowledge is therefore a commodity manufactured, stored, and exchanged by individuals and their organizations. Even knowledge that has been received from other individuals can only be added to the inventory of the mind through the filtering of the clustering process. Our mind does not simply absorb knowledge by, for example, comparing it with established concepts that are stored in the mind. We convert the knowledge thus received into the same clustering process we apply with sensorial inputs. A good allegory is the manner in which we ingest and absorb food, by breaking it down into components that are subsequently assembled into complex substances for use by the body.

Knowledge is the measure of life, but the issues of the levels of knowledge we possess and the *value* of life are a different area of discourse. Lack of a certain level of knowledge in an individual is not an indicator of a lesser life, just as the individual without measurable brain activity is not, by definition, lifeless. As long as the brain exists, whatever its level of functioning, it is a plant that produces and may produce knowledge at any moment. Active brain waves are an indication of brain activity, hence of production and manipulation of knowledge. The value of an individual with minimal brain activity is not different from one with average or above average activity. Inequities in values are of concern to societies and ethicists within them. The precondition for life remains the actual ability to manufacture and manipulate knowledge (at whatever level). Although I compare the human existence to the manufacturing function in organizations, we do not attach less value to individuals as we would close down plants that are not profitable or not producing at a desired level.

Knowledge and Truth

How do we know that our knowledge about ourselves and the world around us is true, and does it matter? If knowledge is the clustering of sensorial inputs, our senses are fallible instruments for scanning the world. They may be, and sometimes are, poorly tuned, insufficiently accurate, and for whatever reason, unable to capture signals, and they are a function of the effort we put into this activity (we may be tired, uninterested, or even hostile in the scanning effort). Fallibility may also occur in the effort to cluster these inputs. We may do so with personal bias due, for instance, to our existing stock of knowledge about the sensorial input, or bias based on history, culture, and religious beliefs that make up our inventory of knowledge or lack the capacity to cluster effectively.

It seems that we have two issues to consider in the link between knowledge and truth. The first is the shortcomings listed above: how we scan and how we cluster. The second is the issue of whether there is objective truth at all. Even scientists sometimes use the terms *knowledge* and *truth* interchangeably. The following statement is attributed to Sir Isaac Newton: "I seem to have been only like a boy playing on the seashore…while the great ocean of truth lay undiscovered before me." But three centuries later, Sir Karl Popper wrote: "The working view of science betrays itself in the craving to be right, for it is not his *possession* of knowledge, of irrefutable truth, that makes the man of science, but his persistent and recklessly critical *quest* for truth" (Popper, 2002). Popper added: "Science never pursues the illusory aim of making its answers final or even probable." All knowledge is not scientific knowledge (or knowledge that is the result of investigation with the scientific method), but Popper's description of the quest for knowledge rather than its possession as truth is in agreement with the definition of knowledge in this book.

Whatever "truth" we assign to knowledge that we share as a community of science or a social group, we do so by consensus, by agreement that is based on whatever criteria, and where the knowledge is collected with whatever method, such knowledge is "true." We may change our mind sooner or later, but at any given time there is a body of knowledge that we accept as the "resident truth" until otherwise contested and a new agreement is drawn.

Is our knowledge a social convention? Knowledge is always the result of the clustering of sensorial inputs and the reliance on the catalog of knowledge already accumulated. Whenever such knowledge is shared with others, it ceases to be a personal affair and is lacking the entrenched mechanisms of quality control and trust that we have with our own senses and the clustering ability. Therefore, the protocol we establish to verify the "truth" of shared knowledge would require reexamination of such knowledge via reclustering. Until then, shared (or explicit) knowledge is merely a *claim* made by others, however trustworthy they may be.

To do so with each and every item of shared knowledge would overly engage mental and cognitive resources that are badly needed to produce "original" knowledge. Imagine a manufacturing plant testing each and every item of material it receives for quality control purposes. Such a plant would use statistical methods to test only samples, and have trust in the vendors and the remaining untested volume of inputs. Our brain is not able to adequately sort through samples of knowledge we constantly receive. Testing *must* be applied to *all* inputs. But the pressures of the struggle for survival are such that a better solution had to be found. This is the *community solution,* by which an agreement is developed where we inherently trust the knowledge we receive as a truthful representation of the world. This convention or agreement is a mechanism that helps us dispense with clustering every piece of knowledge and to accept—albeit temporarily—shared knowledge as "true."

It is in the interest of the individual and the community to establish criteria for agreement on provisional truth of shared knowledge. This frees up cognitive resources to create new knowledge and to arrange shared knowledge we receive within the inventory we already possess. The *community solution* is an instrument that multiplies the power of shared knowledge by diffusing it in a very efficient manner.

As an example, scientific knowledge is shared within the community of science via publications in the academic literature. The criterion for agreement on "provisional truth" of claims and findings thus reported is "peer review." A small group of scholars with background and experience—as well as knowledge in the topic—examines the claims and the results and has the power (with the editors of the journal) to determine acceptability or, for all practical purposes, the "truth" of the material. Other scientists who now share this knowledge usually will not reexamine and recluster this knowledge.

In a world where knowledge *is a subjective* creation by the individual, and it is so crucial to the Darwinian fitness of individuals and communities, truth is a matter for compromises by social conventions, acceptance mechanisms, and a large measure of trust in other members of the community.[15]

But there is still the issue of the objective truth: Can we ever "truly" know our world? Are there what Kant called the things in themselves, namely, reality that is detached from and exists apart from us, the knower? This issue evokes some interesting paradoxes. Frederic Fitch had argued in his paradox of knowability[16] that there is an unknowable truth, thus defying the assertion that "all truths are knowable."

I add the following query: "If all knowledge is subjective (in the sense that it is the product of individual clustering of the individual's sensory inputs), how can it then be true (or truthfully represent the real world)?" For such truth to exist, it would mean "transcending" the knower into the world, so he is not only scanning the world with his senses, but ontologically participating in it, so as to have "first-hand" verification of the true character of the world.[17]

Philosophers, ethicists, scholars, and lay people are all engaged in a constant search for the "truth." Lyotard, for example, had complained that modernity deprived us of the search for knowledge as an end-in-itself in what he called the mechanization of knowledge. In this post-modern environment, knowledge is a commodity that is produced in order to be transacted and exchanged because it is the key component of economic activity and the new driver of production. In this scenario, truth is bended to accommodate the value to producers and to users. The search is not for the "truth" but to what is accepted as truth, as long as it allows for the uninterrupted flow of economic and social exchange.[18]

Scio orbis, non scio veritas—"I know the world, not the truth." To the individual knower all knowledge about himself and the world around him is true knowledge. The knower collects and stores in the knowledge inventory in the mind all knowledge, whether it is "objectively true" or "not true." To know is not to ascertain the truth or to "objectively know" what is true and what is false (namely, how the world "truly" works), but to ascertain that we clustered our sensorial inputs. This means that whenever we cluster sensorial inputs and create and store knowledge, we are *attempting* to ascertain how the world works.

I may even exercise the possibility that we collect all knowledge, true and false, as a mechanism that allows us to play the survival game in the evolutionary struggle. If we do not create false knowledge, how can we tell (or know) that such knowledge is false. In the game of trial-and-error, we invoke from memory knowledge that a certain predator can fly—perhaps we saw a painting on the walls of our cave or this piece of knowledge had been narrated to us by elders. This knowledge prevented us from venturing too far from the cave to greener pastures. Once we do venture, we find that the predator is not as threatening as we believed and we readjust our knowledge by reclustering newly received sensorial inputs. By virtue of the model of the progress of knowledge, continuous cumulation is the means by which we adjust and readjust not only the architecture of our inventory of knowledge but also the truthfulness of its items.

Knowledge Per Se and Knowledge Per Usum

Psychologists and cognitive scientists have suggested that the individual's perception of reality is to the individual the reality with which the individual interacts, and that influences his decisions, behavior, and his dealings with himself and his world. It is only when sharing knowledge with others that "truth" about what is reality is called into question and requires an agreed-upon convention, as well as a methodology for verification that will satisfy all parties. I therefore reject the claim that knowledge which is not shared is not knowledge. Tacit knowledge, embedded in the knower and not shared with others, is simply a different type of knowledge, perhaps the only kind of knowledge.

Instead of the distinction between *tacit* and *explicit* knowledge, I propose that we classify the individual knower's knowledge as: (1) knowledge *per se* (KPS) and (2) knowledge *per usum* (KPU). The classification principle is changed from *where* the knowledge can be located to what the knowledge is for, what its function is, and why we cluster it in the first place.

Knowledge *per se* (KPS) is the knowledge embedded in the knower's mind and to the knower, the truth about himself and the world around him. This is the pool of knowledge from which the knower may draw to share with others and to diffuse in the community. Until and unless convinced by agreement that his knowledge item is not true, all KPS is true to the knower.

Knowledge *per usum* (KPU) is the portion of KPS that we utilize in order to function in the world. It is still a reflection of the world as we got to know it by clustering our sensorial inputs, but it contains knowledge we can equate with specialized skills, such as *procedural* knowledge.[19] In work organizations this type of knowledge is essential for the functioning and success of the organization. Knowledge systems are aimed at capturing and diffusing such knowledge: how to do what, when, and with what. But KPU is more than just skills. It is the mix of skills and experience that guides the knower in the crucial task of survival in the world by knowing the world around the knower and how to perform in it. Knowledge *per usum* draws from knowledge *per se* and is based on the principle that the knower knows the true nature of the world. That is, to the knower, his knowledge of how to function in the world depends on his belief that he knows how the world itself works. Moreover, knowledge *per usum* may be shared, or it may, and usually does, remain embedded in the knower. Indeed, *only* KPU can be shared, whereas KPS remains embedded in the knower and cannot be diffused. KPS is the knower's view of the world or his deep understanding of how the world works and what *is* reality—as he knows it—or the true universe in which he finds himself and in which he must function.

The Good, the Bad, and the Virtue of the Japanese Concept of Ba for Knowledge Creation

In the 1990s several Japanese scholars, led by Ikujiro Nonaka and Hirotaka Takeuchi, introduced to the non-Japanese world of business and scholarship the concept of Ba as a framework for the creation and diffusion of knowledge in work organizations (see Nonaka & Hirotaka, 1995; Nonaka & Noboro, 1998). The concept of Ba is a unique Japanese notion of "a physical, virtual, and/or mental space shared by two or more individuals or organizations" (Nonaka & Nishiguchi, 2001, p. 4). Although not easily translated, Ba is akin to the meeting of the minds within a cooperative engagement, facilitated by or environmentally supportive of collaboration and sharing.

Nonaka and Nishiguchi (2001, p. 14) defined knowledge as "a dynamic human process of justifying personal belief toward the truth." They also argued that the interaction of tacit and explicit knowledge and their conversion from one to the other are the pillars of knowledge creation. They proposed four modes of such conversion: *socialization* (from tacit to tacit), *externalization* (from tacit to explicit), *combination* (from explicit to explicit), and *internalization* (from explicit to tacit). In this model they introduced the notion of the Ba as a platform or place, or the interaction itself between individuals who create and exchange knowledge. The notion of Ba requires that individuals who partake in it must be thoroughly involved. They also argued that "knowledge is embedded in Ba, where it is then acquired through one's own experience or reflections on the experiences of others" ((Nonaka & Nishiguchi, 2001, p. 19).[20]

The good about Ba as the platform or framework that facilitates knowledge creation and diffusion is the mix of Nonaka's description of the environment of Japanese organizations and the emphasis on the role of interactions in the diffusion of knowledge. Nonaka and his colleagues "translated" for the non-Japanese way of thinking the seemingly effective mode of sharing of knowledge that characterized Japanese business companies. They also simplified this process by describing its principles and the modes by which sharing occurs in this peculiar environment. At the same time, they also focused on interaction and sharing, on mentoring and learning that is so well practiced by Japanese managers and envied by their non-Japanese colleagues and competitors.

The bad about Ba is a combination of factors. The concept is unique to Japanese society and to its work environment. It requires homogeneity in the population of managers and workers, and a social system that not only favors a climate of harmony, but demands it. Secondly, the concept of Ba and the process by which knowledge is created and diffused calls for an unbridled desire to participate in the common exchange of what one knows and the blurring of the boundaries between the individual and the collective. Nonaka and his colleagues described this as a "self-transcending process" of interaction, transfer, cooperation, and sharing. Finally, the notion of Ba is a concept that attempts to capture a way of life rather than simply an organizational process.

Because of these factors, it is very difficult to emulate, let alone duplicate, the Ba framework outside the Japanese environment. In our studies of knowledge creation and diffusion in American organizations, we have identified a myriad of barriers to the sharing of knowledge among managers (Rubenstein & Geisler, 2003). To the non-Japanese organization, the notion of Ba and its mode of knowledge creation and diffusion are merely examples of an effective mode (especially of sharing and exchange) of dealing with organizational knowledge, which seems to work in tightly knit social environments.

Ba is of little use to American managers, except perhaps as a means to better understand their Japanese business partners and competitors. In the American and perhaps in all Western business environments, there is more focus on individual competition and the perceived need to retain, not share, knowledge about one's job and one's experience. Knowledge is considered a source of *power.* Its value remains as long as it is *not* shared—like a cherished secret. Once the manager reveals what she knows, everyone in the corporation now possesses the same knowledge and it has lost its competitive value. Everyone in the American or other non-Japanese corporation is a potential (if not actual) competitor for resources, executive recognition, and holding on to one's position and one's job.

The World, Truth, and What We Know

The quintessential problem faced by philosophy has been: Is there a "real" world out there and can we ever "truly" know it? There must be a real world out there or we would not be able to *scan* it with our senses and arrive at some agreement on what this world is like. Because we have different modes of clustering, we must reach an agreement. *Scanning* the world and *clustering* what we scan are different processes. We may receive similar inputs from our senses, but we process them differently, hence arriving at different pictures of the world. The differences are not very striking so they allow for compromise, exchange, and consensus.

There must be a true or real universe out there or we would be unable to receive sensorial inputs that are similar, therefore conducive to consensus. Nevertheless, we never know the "true" world. Our sensors with which we scan this world are imperfect, as is the clustering process. Does it matter? As long as we agree that what we know will always be partial, incomplete, and biased, we also agree that the knowledge we do possess is sufficient or adequate for us to function and to survive. That is, because we are human, we must know or we shall perish.

Logically, then, it is possible to know that there is a true universe, but we only know a biased version of it, not the "true" version of what it is and how it functions. Are we cursed to being always in a mode of "approximate" knowledge, doomed to never "really" know our world?

Fitch's paradox of knowability is relevant to the question above. He argued that if there is an unknown truth, that it is unknown is unknowable. We cannot know what we do not know, so all truths are not knowable. The paradox has been challenged on various grounds, but without going into formal logic, we have argued in this book that by not being omniscient, humans have only incomplete knowledge. I may add that the proposition "all truths are knowable" as the benchmark of knowledge is itself deficient. As non-omniscient, we cannot assert that *all* truths or *all* of anything are whatever attribute we assign to them because we do not know, and will not know, *all* of any kind. So, within the boundaries of the *human* universe, scanable by human senses, we can claim that all that is knowable—by our sensorial inputs and clustering—can be known. However, the argument made above, that to know is not necessarily to know the truth, applies here as well. Whatever we do know does not mean that it is the truth, even in the incomplete universe we are able to scan.

Therefore, because human knowledge is the accumulation of the clustering of sensorial inputs (not the "justified true belief" as some have defined it), we can state that there is a "true" universe out there, existing independently of our senses and of our experience, but knowable truthfully only incompletely. The "true universe" exists as a common platform on which different humans agree regarding similar clusterings of sensorial inputs from this platform. There is an objective external universe, but all human knowledge is subjective, amenable to sharing and diffusion because of interpersonal similarities in knowledge about this "true" universe.

The Principle of Knowledge Adjustment and the Paradox of Post-Modernity

In the context of the scenario I described above, we have imperfect knowledge representing selected slices of the real universe that we are able to scan with our senses. But, in the evolutionary struggle for survival, it is my contention that we are guided by the *principle of knowledge adjustment.* In response to the pressures of such struggle and the need to improve fitness in this arena, human beings have adjusted their knowledge from *intuition* to *cognition.* This shift is the essence of continuous cumulation as the guiding principle of the progress of knowledge.

Intuition is a variant of knowledge *per usum.* It refers to a type of knowledge specific to a well-defined situation. Organization and decision scientists call it "programmed" or "routine" decisions, which are made when the parameters of the situation are known and there is a degree of certainty in the alternatives available to the decision maker and the potential outcomes from these alternatives. This also means that there is a stock of knowledge related to such a situation and that there is no immediate need to collect additional knowledge.

In their progress along the evolutionary scale, when there occurred the increase in brain size, humans shifted their capability to the creation of *cognitive* knowledge. The adjustment of the new type of knowledge creation is the emergence of knowledge *per se,* whereby a much wider variety of knowledge is generated and stored in the human brain, much of it without an immediate utility.

As the human brain was now capable of generating and handling a much larger collection of knowledge, this fact now opened an enormous world of challenges and opportunities. Humans adjusted their brain power to become more cognitive, so they could cluster sensorial inputs in a way that allowed them to generate higher-order constructs about the universe around them that would apply to potential, rather than immediate, situations. This adjustment allowed human beings to think beyond dealing with immediate threats or opportunities, to plan beyond climatic calamities (such as the harshness of winters), and to better confront uncertainty to the point of understanding the workings of their universe and the patterns embedded in seemingly unrelated events.

This much improved ability also facilitated the formation and successful maintenance of social organizations. Cognition permitted shared mental models among members of the social group (e.g., Lupia, McCubbins, & Popkin, 2000). Shared mental models are possible not only as an argument for the existence of a "real" universe, but also as a strong factor in the cohesion of social organizations with shared mental models. Members of these organizations are able to tackle more complex problems that require cognitive abilities and knowledge beyond intuition, and also beyond the purely deductive rationality mode of decision making (see Fiske & Taylor, 1991; Clark, 1997).

To reiterate, knowledge is not evolutionary. It is the human ability to cluster sensorial inputs and to interpret them that has evolved with the appearance of larger human brains.[21] With this ability humans could now target the generation of knowledge toward broader applications. This newly invigorated body of knowledge allows us to formulate hypotheses about the universe and to test new knowledge. The mental framework is not made of a priori concepts, but of constructs derived from sensorial inputs, where these constructs are constantly reinterpreted, or we might say, updated. There are no ingrained or "eternal" truths, just imperfect knowledge about selected aspects of ourselves and our universe that we constantly update. We can do this because of the adjustment from intuition to cognition.

What are the implications for managers and for knowledge systems in organizations? Since all knowledge is subjective but we share mental models as a tool in societal existence, we need to revamp the knowledge systems to allow for agreement and shared knowledge. We must create a "true" universe that would allow members of the organization a measure of agreement. A good example is the strategic vision of the organization and the subsequent "buy-in" of managers in this vision: what the organization stands for and where it is heading. It could be defined as a Western-style Ba, without the social attributes of the unique Japanese climate.

The agreed-upon universe fosters a common ground among managers. A variety of drivers can be applied to promote agreement and a shared perspective. Danger, threats, or a basket of incentives are all potential factors that may encourage sharing. Senior executives must be aware of the fact that sharing knowledge is possible—to a limited extent—and that resistance by managers to participate in knowledge systems can be overcome to a degree.

The principle of adjustment I described earlier has come under duress when we entered the contemporary era of post-modernism. Specialization and the crucial role of knowledge in social and economic activities have engendered a reverse shift toward knowledge *per usum.* Post-modernism is characterized by preference for specialized and relevant knowledge at the expense of knowledge *per se.* As it turns out, we are currently in a state where we know a lot more about a lot less. We have not lost the ability to create knowledge *per se,* but our main preoccupation and cognitive effort are in creating *useful* knowledge.

This state of affairs has the making of the *paradox of modernity.* The evolutionary trajectory is the shift from knowledge *per usum* to knowledge *per se.* But to survive in the post-modern age, we encourage the shift back to knowledge *per usum* for the purpose of survival in these times. Such a shift negates the evolutionary direction.

Perhaps there is a possible resolution to the paradox by distinguishing between the individual and the social group. For the individual there is indeed a shift back toward specialized knowledge. As such, the individual accumulates only the knowledge needed for specialized and well-defined situations in the individual's post-modern universe. It is similar to revisiting the powers of intuitive reasoning. The fitness of the individual in the struggle for survival is segmented and narrowly targeted toward partial sufficiency, rather than self-sufficiency, in knowledge creation, cumulation, and usage.

However, in the *social* context, post-modernism is based on the accumulation of knowledge *per usum* as well as knowledge *per se.* Due to the fact that individuals—as *members of the social entity*—are able to share mental models about the universe, the social entity can act as a repository of both types of knowledge. It thus compensates for the counter-evolutionary shift described in the paradox. In effect, individuals in the post-modern society have abrogated the previous gift that evolution has given them (of being able to create and store knowledge *per se*) for the privilege of existing and surviving in the post-modern society. This is the "price of admission" to this type of social existence—in terms of knowledge creation and usage.[22]

Perfect Knowledge and the Existence of a Supreme Being

I have already established that human knowledge of the universe is imperfect. Throughout history scholars had associated perfect or complete knowledge with the exclusive powers of deities. The ability to know all, from the beginning to the end of time and of the universe's existence, has been duly reserved to the supreme being. Humans were allowed just a "bite of the apple," not ownership of the entire tree of knowledge.

Human beings are merely an ambulatory plant that manufactures knowledge and contains an inventory of knowledge accumulated for eventual use. With this in mind, the human conception of deities is in itself a mark of incomplete knowledge, fear of the unknown and the uncertain—and a cognitive extension of having *some* knowledge to the possibility of *complete* knowledge.

It is a logical extension of having some knowledge to extrapolate the notion that the ability to cluster sensorial inputs is progressive. As clustering progresses, there will be a point where clustering achieves the ultimate step—of having *all* available inputs clustered into meaningful knowledge that can be stored. This is the ability of our omniscient being, namely, a supreme being.

Therefore, if human beings are able to cluster and create some knowledge, there must be other knowledge that exists beyond human reach.[23] Such knowledge will be known by someone other than human beings, therefore one must consent to the existence of an omniscient being.[24]

Even if we add up the incomplete knowledge of all the human beings who ever lived and those who will ever live, we will still be far from complete knowledge. The addition of partial knowledge continues to be incomplete. Complete knowledge is not made of parts. It extends to all time and over all universes and has two attributes of a supreme being. The first is that the possession of complete knowledge allows for pre-planning and, second, compete knowledge *is* knowledge of the "true" universe or of reality in itself. It is being part of reality, enmeshed in it, with no need to represent it: hence the attribute of the supreme being of omnipresence—existing simultaneously everywhere in the universe through knowledge.

Consenting to the existence of a supreme being who is omniscient and omnipresent is not a religious belief but an extension of the perspective of knowledge. Human beings, as knowledge producers, could not logically exist without a parallel or complementary supreme being who knows all.[25]

Knowledge and Ethics

For several millennia, philosophers have developed theories and models of ethical beliefs and ethical behavior. I will not engage the reader with definitions of ethics or the plethora of ethical philosophies. Rather, I will attempt to offer my view of ethics in human interaction seen through the prism of knowledge.

Ethics, as a system of beliefs of what is permissible, acceptable, and non-harmful interaction with others, is based on knowledge. This is not new. The Greek philosopher Plato had argued that as rational beings, we make "right" decisions through rational thinking, and that such decisions reflect what is "virtue" or the right thing to do because they are rational. He who has the knowledge of rational thinking will clearly find that living life according to rational thinking of an enlightened person is also the good, ideal, and happy way of life (e.g., Detel, 2005; Prior, 1991).

The system of human ethics as modeled in this book is based on how this system is created and structured. I am not concerned here with its objectives, its outcomes in social intercourse, nor with its attributes of virtue and righteousness. The model is very narrowly proposed as a model of creation and conception of what is ethical in the human view of oneself and others.

Ethical beliefs and behavior are therefore not about what is "right" or logical, but how they are forged from concepts which are the clustering of our knowledge. Ethics is utilitarian because of the manner in which knowledge is structured and progresses. Concepts of ethics are based on human knowledge and experience.

Consider, for example, the concept of *justice* as an illustration of what moral concepts represent in such a scheme of ethics.[26] Humans do not possess the high-order notion of justice as an a priori notion embedded in their cognition. Rather, as knowledge about the world is continually accumulated by clustering of sensorial inputs, we build a reservoir or collection of events or instances in which sensorial inputs had been clustered into the experience that certain acts done to us or that we do to others are harmful. As such, knowledge accumulates and is shared in the social environment. We reach a consensus with others that certain behavior is harmful and should be avoided, or in cases where it does occur, to be punished by the social order.

Therefore, we do not look for justice for the sake of the concept itself, but because we possess the knowledge that such behavior is positive and rewarding and the opposite is harmful (to us and to others).[27] But, clearly the possession of knowledge does not guarantee the making of ethical choices. There is not a direct causal link between what we know about a concept and decisions we make in our lives. We may possess knowledge about justice or the value of a human life and make decisions and behave in a manner that negates this knowledge and causes injustice and even the taking of a human life.

As knowledge in the form of notions and concepts is accumulated by individuals, they share some of it with others upon the creation of social groups and organizations. There is a transformation in these concepts (such as "justice") whereby they become the property of the collective. A concept such as justice is then accepted by the society as a principle of behavior due to the consensus and agreement on the part of the members of the society. The concept remains an individual experience and depends on individual knowledge.[28] As all knowledge is individualized, so is ethics.[29]

The driving force of the emergence of ethical principles and ethical considerations is the role that knowledge plays in the self-preservation and the survival of humans. We move from sensorial inputs to clustering knowledge to forming higher-order constructs to ethics. In this respect, ethics is the product of experience and is utilitarian. Individual knowledge about behavior and its consequences leads to a mechanism of selecting the modes of behavior (toward self and others) which assists in minimizing harm and in promoting self-preservation. This includes behavior that is altruistic. The sacrifice of one's health, wealth, life, and limb for family and others is a form of preservation of one's being and future, and the exercise of knowledge about the value of family and other humans in one's existence.[30]

The view of ethics as knowledge also suggests that to know that God exists is not to know what God wants or wishes believers to do and how to behave. The existence of human knowledge conforms to the notion of a supreme being, but one cannot extrapolate from this notion to a set of ethical principles. Only this supreme being understands concepts that transcend experience. Human beings in their frailty and their limited physical and mental capabilities can only conceive those conceptual notions that are generated by clustering of sensorial inputs. Any attribution of ethical concepts to the dictates of the omnipotent supreme being is an artificial mode of validating such concepts and making them socially palatable and easier to be accepted. Rather than promoting these concepts as ideas of other humans, the divine or ethereal attribution enshrines them with the unquestionable qualities of eternal truism.[31]

Knowledge, Technology, and Society

Because of the limitations of human capabilities, knowledge that humans possess is the progenitor of technology. The creation of technological artifacts and tools is an extension of human senses. If I cannot see far enough, I will develop an artifact such as a telescope that allows me to scan the heavens. If my sense of smell is relatively poor, I will develop a tool that can "sniff" explosives and identify them before they can be loaded on commercial airplanes.

As we humans develop these tools and instruments that extend our capabilities and our senses, the handling of such technology becomes increasingly specialized and sophisticated. Few of us know how this technology works and how to use it. This generates a fear that knowledge unleashes a threat of disaster on unsuspecting people and their organizations.[32] Developments in biotechnology, genetic engineering, and nanotechnology are considered to be of imminent risk as they become so complex and pervasive that even their handlers will soon lose control over them. These fears are expressed not only by Luddites (people whose livelihood is threatened by new machines and new technology—after the 1811 destruction by angry workers of factory machines in the United Kingdom), but also by respectable scientists.

Although such concerns are justified, the answer is not to curb or restrict or severely regulate the production of knowledge and the development of new technologies. On the contrary, the answer is to counter these threats with knowledge. It is impossible to "put the genie back in the bottle." Once knowledge is acquired and a technology is engendered, they are an integral component of the human catalog of what we know and who we are.

To prevent possible catastrophes from misuse of technology, we need to employ bright and talented minds to develop countermeasures to such potential disastrous mishaps. To paraphrase an old military saying: "All that is built by the human mind can be defeated or controlled by the human mind."[33]

How can knowledge be captured and diffused in social organizations? It is quite possible to imagine social organizations as an extension of individuals and to imagine that the sum total of the knowledge in organizations is synergetically multiplied. I do not believe in the reification of organizations. As instruments for the diffusion of knowledge from individuals to other individuals, organizations are merely the platform for such exchanges.

Therefore, I do not believe that we can have such phenomena as *organizational learning* or an *organizational knowledge base*. Social and work organizations do not know nor do they learn. Only individuals know and only individuals learn. We collect individuals' knowledge for sharing and diffusion with other individuals, but this does not mean that the collection of knowledge has a life of its own or that it is greater than the sum of its parts.

To reiterate, the design of knowledge systems in organizations is the putting together of the collection of knowledge accumulated by individuals who are or had been members of the organization or who had ties and relationships with it (such as vendors, competitors, and regulators). Such knowledge systems are not the repository of the "wisdom" of the organization, but they do offer a substitute of the "eyes and ears" of the organization.[34] In effect, such systems tell us "what we know about a given event or situation."

Similarly, the collection of knowledge in social organizations is not the collective wisdom of the society. As the knowledge in the organization is embedded in both its records and in people's minds, the totality of knowledge in the society would be the collection of *all* research and of *all* the people who have knowledge for these organizations.[35] Currently this is not possible to compute.

Yet, with an increasing effort to share and exchange knowledge that is so pervasive in today's social organizations, we may be able to treat such shared knowledge as a commodity. Balancing and regulating this commodity within society would require a regimen of rules and allocation criteria similar to that used for products and services and other forms of distribution and control of wealth. Even without the established ability to measure such knowledge commodity beyond explicit nuggets found in records, we may be able to extend the rules used for intellectual property to all forms of knowledge.[36]

References

Clark, A. (1997). *Being there: Putting brain, body, and world together again.* Cambridge, MA: MIT Press.

Detel, W. (2005). *Foucault and classical antiquity: Power, ethics, and knowledge.* New York: Cambridge University Press.

Fiske, S., & Taylor, S. (1991). *Social cognition.* New York: McGraw Hill.

Kennerknecht, I., et al. (2006). First report of non-syndromatic hereditary prosopagnosia (HPA). *American Journal of Medical Genetics, Part A*(June 30).

Lupia, A., McCubbins, M., & Popkin, S. (Eds.). (2000). *Elements of reason: Cognition, choice, and the bounds of rationality*. New York: Cambridge University Press.

Luria, A. (1988). *The mind of a mnemonist*. Cambridge, MA: Harvard University Press.

Nonaka, I., & Hirotaka, T. (1995). *The knowledge creating company*. New York: Oxford University Press.

Nonaka, I., & Nishiguchi, T. (2001). *Knowledge emergence*. New York: Oxford University Press.

Nonaka, I., & Noboro, K. (1998). The concept of Ba: Building a foundation for knowledge creation. *California Management Review, 40*(3), 1-15.

Popper, K. (2002). *The logic of scientific discovery* (new ed.). Abington, UK: Routledge.

Prior, W. (1991). *Virtue and knowledge: An introduction to ancient Greek ethics*. Abingdon, UK: Routledge.

Rubenstein, A.H., & Geisler, E. (2003). *Installing and managing workable knowledge management systems*. Westport, CT: Praeger.

Endnotes

[1] This term describes a person with an usually prodigious memory: named after Mnemosyne, the Greek goddess of memory. The book is: Luria, A. (1987). *The mind of a mnemonist*. Cambridge, MA: Harvard University Press.

[2] See, for example, Grossberg, S., & Gutowski, W. (1987). Neural dynamics of decision-making under risk: Affective balance and cognitive-emotional interactions. *Psychological Review, 94*(3), 300-318. The authors concluded that in framing alternative decisions, humans may not choose to maximize their value with the decision, rather to balance their emotional status. In framing decisions under risk, individuals create knowledge and by doing so may trigger complex mechanisms of clustering that also include cognitive-emotional interactions. Also see: Grossberg, S. (1982). *Studies of mind and brain: Neural principles of learning, perception, development, cognition, and motor control*. Boston: Reidel Press.

3   Philosophers and linguists will probably argue about the domain of definition and its distinct action as opposed to explanation. In my view, the lens of knowledge is a means, not *the* means, of looking at the world. But by doing so we are able to apply this lens to virtually every aspect of life and existence, similar to how "string theory" is applied (at least in theory) to every item in the physical world.

4   Socrates did not produce any written legacy of his work. We know about his ideas through the *dialogues* written by his pupil Plato (427-347 BC).

5   This school of philosophy was founded by Soren Kierkegaard (1813-1855), the Danish scholar who considered human existence as a constant struggle in which individuals are left to their devices and must make hard choices in every eventuality they encounter. His ideas were refined and expanded in Sartre's philosophy.

6   Sources: Sartre's lecture given in 1946: "Existentialism is a Humanism"; and Sartre, J. (1977). *Being and nothingness.* New York: Washington Square Press.

7   Phenomenologists, such as Husserl and Heidegger, would argue that knowledge does exist outside the real world and the boundaries of experience. I am suggesting that knowledge is a reflection of experience and reality in the sense of the clustering of sensorial inputs, and at the same time a tool in manipulating experience. The reader who is familiar with Heisenberg's theories will find an analogy with a system of measurement that measures phenomena in the physical world, but also influences the phenomenon and the measures thus obtained by the mere fact that a measurement effort is attempted at all.

8   These questions are further explored below in the epilogue "Toward a Theory of Knowledge."

9   This also includes knowledge embedded in genetic material because such material existed already in the parent's being, before its transfer to the descendants.

10   Not exactly the distinction between *explicit* and *tacit* knowledge because the latter always has the potential of becoming shared and diffused.

11   This does not necessarily imply that reality is what we perceive it to be, nor that "things-by-themselves" or real things exist outside the knower. I am only saying that what we know and how much we know is not indicative of the "true" existence of the physical world. We simply cre-

ate knowledge by clustering sensorial inputs, and by doing so we are not justifying nor asserting the true existence of that which our senses had captured.

[12] Oh, no! I am joining in the same sentence the Biblical account and the tenets of evolution. Perhaps my ideas put forth in this book would serve as a bridge in this dispute.

[13] See Dorus, S. et al. (2004). Accelerated evolution of nervous system genes in the origin of *Homo sapiens. Cell, 119*(December 24), 1027-1040; and Stedman, H. et al. (2004). Myosin gene mutation correlates with anatomical changes in the human lineage. *Nature, 428*(6981), 415-418. The authors concluded: "This represents the first proteomic distinction between humans and chimpanzees that can be correlated with a traceable anatomic imprint in the fossil record" (p. 415). This study suggests that changes in masticatory muscles in humans appeared about the same time as changes in the size of their brains, all this upon the divergence of humans from chimpanzees. However this happened and whichever mutation came first (muscles or size of the brain), humans were endowed at a given point on the evolutionary ladder with the ability to produce knowledge. Such a coincidence was necessary because the mutation in the muscles of the mouth would have made humans less able to compete in their environment. The increase in the size of brain (hence added ability to produce knowledge) probably compensated for the loss of muscle and even catapulted humans to new highs on the evolutionary scale.

[14] Perhaps one can designate this perspective of humans as a *mechanistic* approach. It lacks a spiritual coating, but it is not an extension of existential beliefs. Rather, as I am a product of the latter part of the twentieth and the early years of the twenty-first centuries, I view knowledge as both a commodity and the essential component of the human evolutionary progression.

[15] In extreme examples, we do not second guess instructions and recommendations from our physicians, lawyers, and accountants. There is a measure of trust that the knowledge they share with us is indeed a true representation of the world. See: Tallis, R. (2005). *The knowing animal: A philosophical inquiry into knowledge and truth.* Edinburgh: Edinburgh University Press.

[16] See discussion of the paradox in Chapter XIV of this book.

[17] Let me use a cliché such as "walking in someone's shoes" to really appreciate the other person's experiences which thus become *your* experiences, and we therefore maintain the principle of subjectivity. See, for example: Poli, R. (Ed.). (1993). *Consciousness, knowledge, and truth.* New York: Springer-Verlag.

[18] In the commercial world we have seen various applications of social conventions of the "truth." In 2002 in response to business accounting scandals, the United States Congress enacted legislation proposed by Senator Paul Sarbanes and Representative Michael Oxley that called for more "truthful" reporting of corporate transactions.

[19] See Chapter IX of this book.

[20] Ba can be of the following types: (1) originating (face-to-face), (2) dialoguing (peer-to-peer), (3) systematizing (on-site collaboration), and (4) exercising (on-site socialization).

[21] In human development, toddlers are only able to identify themselves and create mental models of "self" and "other" after a certain age, usually two years of age. This mental agility requires not only a better developed brain than that of a baby, but also a stock of knowledge that has been accumulated since birth. In fact, humans in their infancy decipher patterns of speech to establish regularities, such as intonations that mean favor or disfavor, menace or acceptance. This ability is also shared by some primates, and studies have found also by rats (see: Toro, J. et al. (2005). Effects of backward speech and speaker variability in language discrimination by rats. *Journal of Experimental Psychology–Animal Behavior Processes, 31*(1), 95-100.

[22] Sociologists and social psychologists of course list other factors in the sacrifices of post-modern existence such as alienation, depression, and other mental and social ills.

[23] Not necessarily "true" knowledge. See, for example, the paradox of knowability.

[24] Such existence is in the limited context of possessing knowledge, not necessarily having unlimited powers. That is, being omniscient, not necessarily omnipotent. One needs to include a causation between the possession of knowledge and the exercise of power. Nor does complete knowledge equate the original design of the universe. Omnipotence is a belief, not knowledge.

25   Personally, I believe in the existence of God beyond the perspective of knowledge. However, it seems to me that it is irrelevant how or why we believe in its existence, as long as we believe. The perspective of knowledge is, in my view, an elegant way of arriving at what religions have always taught. See, for example, Stace, W. (1970). *The theory of knowledge and existence.* Westport, CT: Greenwood Press.

26   The reader may wish to consult the literature on ethics, such as Lloyd, G. (1994). *Port of nature: Self-knowledge in Spinosa's ethics.* Ithaca, NY: Cornell University Press. Also: Johnston, P. (1999). *The contradictions of modern moral philosophy: Ethics after Wittgenstein.* Abingdon, UK: Routledge; and: Rawls, J. (1999). *A theory of justice.* Cambridge, MA: Belknap Press.

27   This may explain why children are not aware of the ethical implications of their behavior, nor are animals other than adult humans. They all lack the knowledge that drives the ethical choices.

28   Social sharing is not a guarantee for ethical behavior. In the animal kingdom there are examples of social organizations, yet the lack of individualized knowledge creation and its progress prevents these animals from generating and exercising what we would consider ethical behavior.

29   I agree here with those philosophers who emphasized individual responsibility for ethical behavior. Other philosophers, such as Kant, had argued that suicide, for example, is unethical because if it is generalized to the entire society, it would mean the end of this society. Individual ethics is explained in terms of its generalization to society. In the model of ethics as knowledge, suicide would be unethical behavior because individuals *know* that such behavior is harmful to one's self and to others (such as family members who are left behind). I reiterate that knowing does not prevent the commission of such unethical acts. To know and to act are two different phenomena. Actions and behavior are influenced by a myriad of factors, in addition to knowledge. I also wish to reemphasize that although in this book I have often used the term "subjective knowledge," a better term is "individualized knowledge." The term "subjective" may have a connotation of knowledge that is internalized or mainly perceptive, with little tendency to be shared with others.

30   This model of ethics clearly leads me to disagree with social philosophers and political scientists such as Karl Marx, George Friedrich, Hegel, and their twentieth-century followers (e.g., Herbert Marcuse) who focused

on the role of society as the driving force of human existence and human ethics. See, for example, Breckman, W. (1999). *Marx, the young Hegelians and the origins of radical social theory: Dethroning the self.* New York: Cambridge University Press.

[31]   Human history is a burgeoning repository of examples of behaviors that were validated or justified with divine providence—later to be reversed and declared wholly unethical. The phenomena of slavery, various forms of discrimination, and wars are salient illustrations. Nonetheless, divine attribution of ethical principles serves a crucial social function. The individual basis of ethics-as-knowledge does not automatically generate ethical *behavior.* Armed with the knowledge to behave ethically, individuals then face the tremendous responsibility of exercising such knowledge through their choice of action. Social leaders find themselves in the tenuous position of guardians of the social order by enforcing the shared knowledge about what is ethical. Here is the essence of the problem in societies. These leaders have to interpret the shared knowledge and impose *their* version of what is ethical. Hence the different ideologies, customs, and ethical norms of behavior among the different social orders and social frameworks.

[32]   I discussed this topic in Geisler, E. (2001). *Creating value with science and technology* (pp. 292-314). Westport, CT: Greenwood.

[33]   This is a social and political issue. There is a need for social leaders to promote mental abilities over physical skills so as to attract the best minds to the service of society.

[34]   I define "wisdom" as the knowledge applied in making choices that are compliant with a given situation ("how to do things right"). This means that wisdom is specific to the factors and parameters of a given circumstance. What may be "wise" in one case may be wholly unwise in another.

[35]   This would be equivalent to the national or domestic gross product (GNP) as a measure of the total value of all goods and services produced in a society in a given period of time. For knowledge, a measure of *gross social knowledge* would be the totality (in some measure of knowledge such as "nuggets") of all knowledge accumulated in all social organizations up to and including a specific point in time. With our current capabilities, this measure is still within the realm of the desired yet impossible aim.

[36]   The reader may continue this line of thinking (which is characteristic of a twenty-first-century person) to the development of the criteria and principles needed to implement such a notion. For example, see: Marlin-Bennet, R. (2004). *Knowledge power: Intellectual property, information, and privacy.* Boulder, CO: Lynne Rienner; and Drahos, P., & Mayne, R. (Eds.). (2002). *Global intellectual property rights: Knowledge, access and development.* New York: Palgrave-MacMillan.

Epilogue

Towards a General Theory of Knowledge

The pursuit of a general theory of knowledge has recently become a hobby for a diversity of scholars working in the area of knowledge and knowledge management systems. I am perhaps an exception because a general theory of knowledge (GTK) was not the main objective of this book. As the various chapters coalesced into a cohesive model of knowledge, the blending of structure and progress evolved into compatible elements of a general theory.

On the road toward a general theory, we encountered several definitions of knowledge and some interesting taxonomies. A working definition of knowledge contends that it is "a fluid mix of framed experience, values, contextual information, and expert insights that provides a framework for evaluating and incorporating new experiences and information. It originates and is applied in the minds of the knowers."[1] This and similar definitions consider knowledge to be both an entity—such as a framework—and a process.

A different approach examines whether knowledge is energy or matter. Is knowledge an ontological entity (does it exist on its own) or is it energy that flows through the labyrinths of the mind?[2] Imagine a single item of knowledge, such as the primary element of the clustering of sensorial inputs—the KANE. Does such a KANE ever cease to exist or, like energy, is it merely transformed into another form by clustering with other KANEs and by assuming the form of a higher-order construct? If knowledge is shared with other individuals and, by convention, entered into records, does it become "matter"?[3]

Knowledge is the texture of our existence. What we know animates our being and endows us with purpose and with the ability to survive. As such, knowledge is both energy and matter. It is energy in the sense that it resides within the mind of the knower and can undergo a series of transformations that changes its format and increases its complexity while keeping its elemental attributes relatively stable. Knowledge is also matter, when the transformation it undergoes includes sharing with others and making the transfer to a mode of interaction *outside* the knower. The only way that knowledge can be shared and manipulated by others is if it is an entity separate from the knower.

If a tree falls in the forest, does it matter? Yes, to the tree and to the fauna and flora that lived on it, in it, and off it, but perhaps not to you and me who reside distances away. Conversely, every knowledge matters. In our model of the progress of knowledge, social existence and the welfare of society depend on continuous cumulation of what we know. A theory of knowledge would provide a perspective of what is knowledge.

A tree falling in the forest is an event in the world we inhabit. If I do not have knowledge of it, the event does not matter to me. A tree falling in the forest may trigger a "butterfly effect" by reducing the green lungs of the globe we inhabit, thus increasing the level of carbon dioxide and elevating the threat to all life on the planet—including mine.

But, all of the arguments above mean that we possess knowledge about the phenomenon, we presume, is created by the falling tree. All knowledge is individual, so that although I am told, or taught, that the phenomenon of the environmental threat because of a fallen tree indeed exists, it is still only "shared knowledge," not *my* knowledge. I may be adequately convinced to an extent that I will contribute to a social endeavor to deal with the threat, but unless I gain knowledge of the event of the fallen tree by clustering my sensorial inputs, I do not know, hence, it does not matter to me that a tree falls in the forest.

Acting within the social context on shared knowledge but not having knowledge through clustering of one's own sensorial inputs is the price we pay for being a member of a social entity and for enjoying its benefits of safety, security, psychological aspects of belonging, and the economic benefits that we accrue from such membership. A theory of knowledge must take these issues into account and explain the apparent dichotomy of individual knowledge and that which we accept as the price of admission to membership in society.

A theory of knowledge would have two key components: how knowledge is generated and structured, and how it grows and progresses. Hypothesis 1 of such a theory would state: "Knowledge is generated when there is clustering of sensorial inputs and the human (or artificial) brain interprets this clustering." Hypothesis 2 would state: "Knowledge progresses and grows through a process of continuous cumulation, whereby added knowledge is joined with the existing knowledge base to provide, when feasible, new architectures and meaning."

In the human or social organization, knowledge systems are created to capture the knowledge that exists in records and in people's minds. The key issues with these systems are: (1) which type of knowledge to capture, (2) how to capture, and (3) how to utilize the knowledge captured within the system. These three issues harbor a host of barriers (technical, organizational, behavioral, and economic) to the successful implementation of knowledge management systems.

Knowledge and Ideas

The link between knowledge, consciousness, and ideas has received considerable attention throughout this book. In the emerging theory of knowledge, consciousness is a state of the operation of the human brain when it is engaged in clustering of sensorial inputs and transforming them into knowledge. To know is to exist, therefore to know is to have consciousness. Comatose patients, for example, have a relatively functioning brain but lack the ability to know—to cluster sensorial inputs and to interpret them within their stock of accumulated knowledge. Their senses may continue to monitor their environment to a certain extent, but the inputs, if any, are not clustered.

A similar view can be applied to the link between knowledge and ideas. I have already rejected earlier the notion that ideas ("memes") can be transmitted and spread like a virus, namely, that these ideas have a life of their own, hence they also have a separate existence. Ideas are knowledge in the form of higher-order constructs. They are a reflection of the clustering of our sensorial inputs and are engendered by such clustering.[4]

What We Know, How We Know, and Why It Matters

What we know is only an outcome of how many inputs our senses have provided and how we manipulated the stock of knowledge we continually accumulate. It matters to us as individuals because knowledge is the lens or prism for every aspect of our existence.

On the other hand, we also exist in social organizations where we interface with others and where we exchange what we know. This provides us with the ability to share and to survive and prosper with the advantages that social organizations offer us as compensation for our frailties in our struggles with our natural environment.

Knowledge, therefore, is not just some tool we pursue as a hobby to enrich our frontiers of imagination and creative desires. Rather, knowledge is *the* essential ingredient of our existence and of our survival. It is the tapestry of our mind and the thread that ties us together with others in our social condition. We must continue to vigorously increase the effort not only to gain more knowledge, but to understand its making, its make-up, and how we may continuously improve our existence with what we describe as "knowledge."

A final word to the reader who relentlessly accompanied me through the intellectual tribulations of this book. My perspective on how knowledge is structured and how it progresses has evolved during the writing of this book. The second half of Chapter XVI was drafted in early January 2005 during a flight on United Airlines from Washington, DC, to Chicago, Illinois. Like a good mystery novel, I did not quite know how this entire scheme would come together, but together it came!

I do not offer here definitive answers to the many current questions about knowledge and its role in our complicated lives. My contribution has been in building a model of how knowledge is generated, how it is structured, and how it grows and progresses. To paraphrase Winston Churchill and Stephen

Jay Gould,[5] this may well be the end of the beginning in our effort to theorize and to model human knowledge. The notions of *continuous cumulation* and *I know, therefore everything* are keystones of how I view knowledge. They have, I believe, important implications for how we design our knowledge bases and knowledge systems. The time has come to revisit our effort in collecting and structuring knowledge systems, and to make them more effective and to shape them in the way we frame the knowledge in our minds.

Endnotes

[1] Some illustrative publications are: Moser, P., Mulder, D., & Trout, J. (1997). *The theory of knowledge: A thematic introduction.* New York: Oxford University Press; and, Lehrer, K. (2000). *Theory of knowledge.* Boulder, CO: Westview Press.

[2] Davenport, T., & Prusak, L. (2000). *Working knowledge* (p. 5). Cambridge, MA: Harvard University Press.

[3] See, for example, Quine, W. (1977). *Ontological reality.* New York: Columbia University Press. Also see: Schlick, M., Blumberg, A., & Feige, H. (1985). *General theory of knowledge.* Peru, IL: Open Court.

[4] The reader may be interested in the work of Hans Vaihinger (1852-1933) and his theory of "fictions." Vaihinger was a Kantian scholar, and his work greatly influenced the American psychologists George Kelly (1905-1967) and Alfred Adler (1870-1937). Vaihinger's "as if" philosophy provided the background for role playing as a tool in psychotherapy. See Vaihinger, H. (1968). *The philosophy of "as if."* New York: Barnes & Noble. Vaihinger said: "The object of the world of ideas as a whole is not the portrayal of reality—this would be an utterly impossible task—but rather to provide us with an instrument for finding our way about more easily in the world" (p. 15). He argued that because we cannot know reality per se, we create "fictional" constructs of the world so we can exist in mental comfort. I arrived at a similar view, although coming from the clustering of knowledge and viewing ideas as the constructs of knowledge.

[5] In the title of his book: Gould, S. (2002). *I have landed: The end of a beginning in natural history.* New York: Harmony Books.

Glossary

A Posteriori Knowledge: From the Latin "which comes after"; describes knowledge that the mind derives by inductive processes from sensorial perceptions and from experience.

A Priori Knowledge: From the Latin "which comes before"; describes knowledge that the mind derives from rational processes of preexisting ideas or concepts, which then allow by deduction to arrive at other forms of knowledge.

Analytic statement: A statement that is true by definition. This is also known as an "uninformative tautology" because the information given by the predicate is already given by the subject.

Architectonic: Kant's term describing the logical structure of a system of philosophy as dictated by human reason.

Architecture: The form or structure of knowledge resulting from accumulation of clustering of items of knowledge.

Categories: Kant's term describing the concepts in the mind, acting as a lens through which objects of the world are examined. There are four such categories: quantity, quality, relation, modality.

Empiricism: Philosophical theory or a school of thought that claims that all true knowledge is based on experience, hence a priori existence of ideas or knowledge (prior to experience) is not legitimate or true knowledge.

Epistemology: A branch of philosophical inquiry concerning the study of the nature of knowledge and its effects on reason and ethical conduct.

Eschatology: A set of beliefs, philosophical explanations, and a school in historical analysis concerning the last chapter of human history, or the "end of the world."

Ethnology: The focus in the study of knowledge on the ethics of human behavior.

Etymology: The systematic study of the origin of language by focusing on the structure of words and their initial diffusion in languages.

Hermeneutic: The systematic study of interpretation (of texts, symbols, etc.).

Homeostasis: A state of equilibrium, relative stability, or balance within different elements of a system or group.

Inductive Argument: An argument whose premises support the conclusion so that if these premises are true, the conclusion is probably also true.

Indexical: Concept or word that is used to represent other concepts or words, in a way that indexes stand for other concepts or words in a specific context in which they are employed.

Intuition: Kant's term describing the ability of the mind to perceive relations among sensorial representations of the empirical world, when these are given in time and space.

Logical Empiricists: A contemporary philosophical school which argues that scientific verification, as required by the logical positivists, is philosophically unverifiable. Hence, this school broke away from the early positivists in the distinction between analytic and synthetic statements.

Logical Positivists: A philosophical school developed in the early twentieth century that argued that scientific verification is key to attaining true knowledge.

Model A: A model of knowledge structure and progress based on transformations of information and databases. Also described as "Primitive Model."

Model B: A model of knowledge structure and progress, proposed in this book, based on the clustering of sensorial inputs as this phenomenon occurs in the human mind. Also described as "Neuronic Model."

Ontology: The branch of philosophy devoted to the study of being, its nature, and the kinds of physical, natural, or human existence.

Paradigm: An undisputed example of an archetype, also used to express an entrenched school of thought or a set of beliefs.

Parallax: The difference in position or direction of an object when observed from two distinct vantage points. May also be used metaphorically to describe the phenomenon of two different observations of the same event.

Phenomenology: A philosophical approach or a school of thought proposed by Edmund Husserl whose focus is on describing and understanding experience as it is perceived by human consciousness, so that the consciousness can refer to the physical world outside itself.

Positivism: A school of thought in the philosophy of knowledge in which knowledge is solely derived from experience and empirical investigations.

Prosopagnosia: An inherited impairment in the recognition of faces of other people, including sometimes the face of the patient. This condition does not preclude the person from recognizing colors, clothing, or emotions. Also known as "face blindness."

Rationalism: A school of thought in the philosophy of knowledge contending that reason and the exercise of logic are solely responsible in the attainment of genuine or true knowledge.

Reductionism: A methodology of research by which the researcher explores increasingly smaller components of the phenomenon.

Reification: The process by which we ascribe material properties to an abstract.

Semantics: The study of the meaning of words, sentences, expressions, and signs of the human language (from the Greek word *semantikos,* meaning significance).

Semiotics: A general theory of signs and symbols, in artificial and natural languages, which includes such areas as syntactics and semantics.

Syllogism: A set of statements usually containing two premises from which a third statement, a conclusion, is deduced.

Synaesthesia: Joining sensations together involuntarily that are usually experienced individually.

Synthetic Statement: A statement whose true value solely depends on experience and empirical observations.

Transcendental Knowledge: Kant's term describing knowledge about the world obtained by pure reason.

Additional Readings

Ackerman, M., Pipek, V., & Wulf, V. (Eds.). (2002). *Sharing expertise: Beyond knowledge management.* Cambridge, MA: MIT Press.

Adler, P.S. (1989). When knowledge is the critical resource, knowledge management is the critical task. *IEEE Transactions on Engineering Management, 36*(2), 87-94.

Agarwal, R., Echambadi, R., Franco, A., & Sarkar, M. (2004). Knowledge transfer through inheritance: Spin-out generation, development, and survival. *Academy of Management Journal, 7*(4), 501-522.

Ahuja, R., Maugnanti, T., & Orlin, J. (1993). *Network flows: Theory, algorithms, and applications.* Upper Saddle River, NJ: Prentice Hall.

Alavi, M., & Tiwana, A. (2002). Knowledge integration in virtual teams: The potential role of knowledge management systems. *Journal of the American Society for Information Science and Technology, 53*(12), 1029-1037.

Albert, S., & Bradley, K. (1997). *Managing knowledge: Experts, agencies, and organizations.* New York: Cambridge University Press.

Alford, R. (1998). *The craft of inquiry: Theories, methods, evidence.* New York: Oxford University Press.

Al-Hawari, M., & Hasan, H. (2004). Knowledge management styles and organizational performance: An empirical study in a K-space framework. *Journal of Information & Knowledge Management, 3*(4), 347-372.

Allee, V. (1997). *The knowledge evolution: Expanding organizational intelligence.* Boston: Butterworth-Heinemann.

Allee, V. (2002). *The future of knowledge: Increasing prosperity through value networks.* New York: Elsevier Science & Technology.

Alvesson, M., & Karreman, D. (2001). Odd couple: Making sense of the curious concept of knowledge management. *Journal of Management Studies, 38*(7), 995-1018.

Amaravadi, C., & Lee, I. (2005). The dimensions of process knowledge. *Knowledge and Process Management, 12*(1), 65-77.

Amin, A., & Cohendent, P. (2004). *Architectures of knowledge: Firms, capabilities, and communities.* New York: Oxford University Press.

Anderson, B., Howells, J., Miles, I., & Roberts, J. (Eds.). (2002). *Knowledge and innovation in the new service economy.* London: Edward Elgar.

Anderson, J. (1983). *The architecture of cognition.* Cambridge, MA: Harvard University Press.

Andriessen, J., & Fahlbruch, B. (Eds.). (2004). *How to manage experience sharing: From organizational surprises to organizational knowledge.* Burlington, MA: Butterworth-Heinemann.

Aquila, R. (1983). *Representational mind: A study of Kant's Theory of Knowledge.* Bloomington: Indiana University Press.

Arbnor, I., & Bjerke, B. (1997). *Methodology for creating business knowledge* (2nd ed.). Thousand Oaks, CA: Sage.

Argote, L. (1999). *Organizational learning: Creating, retaining, and transferring knowledge.* Norwell, MA: Kluwer.

Argyris, C. (1980). *Inner contradictions of rigorous research.* New York: Academic Press.

Argyris, C. (1982). *Learning and action: Individual and organizational.* San Francisco: Jossey-Bass.

Argyris, C. (1993). *Knowledge for action.* San Francisco: Jossey-Bass.

Aronson, J. (1995). *Realism rescued: How scientific progress is possible.* New York: Open Court.

Ashenhurst, R. (1997). Ontological aspects of information modeling. *Minds and Machines, 6*(3), 287-394.

Ashenhurst, R. (2003, April 23). Information modelings: Life begins at 40? *Proceedings of the 40th Anniversary Program of the Chicago Chapter of the ACM.*

Assaro, P. (2000). Transforming society by transforming technology. *Accounting, Management and Information Technologies, 10*(4), 257-290.

Audi, R. (1998). *Epistemology: A contemporary introduction to the theory of knowledge (Routledge contemporary introduction to philosophy).* New York: Routledge.

Awad, E., & Ghaziri, H. (2003). *Knowledge management.* Upper Saddle River, NJ: Prentice Hall.

Ayer, A. (1952). *Language, truth, and logic* (reprint ed.). New York: Peter Smith.

Badaracco, J. (1991). *The knowledge link: How firms compete through strategic alliances.* Boston: Harvard Business School Press.

Baldwin, J., & Hanel, P. (2003). *Innovation and knowledge creation in an open economy.* New York: Cambridge University Press.

Barabasi, A. (2003). *Linked: How everything is connected to everything else and what it means.* Berlin: Plume Books.

Barbour, J. (2000). *The end of time.* Oxford, UK: Oxford University Press.

Bar-Hillel, Y. (1964). *Language and information.* Reading, MA: Addison-Wesley.

Barnes, B., Bloor, D., & Henry, J. (Eds.). (1996). *Scientific knowledge: A Sociological analysis.* Chicago: University of Chicago Press.

Barnett, S., & Ceci, S. (2002). When and where do we apply what we learn? A taxonomy for far transfer. *Psychological Bulletin, 128*(4), 612-639.

Baskerville, R., & Myers, M. (2002). Information systems as a reference discipline. *MIS Quarterly, 26*(1), 1-14.

Baskerville, R., Smithson, S., Ngwenyama, O., & DeGross, J. (Eds.). (1994). *Transforming organizations with information technology.* Amsterdam: North-Holland.

Baumard, P. (1999). *Tacit knowledge in organizations.* San Francisco: Sage.

Beazley, H., Harden, D., & Boenisch, J. (2002). *Continuity management: Preserving corporate knowledge and productivity when employees leave.* New York: John Wiley & Sons.

Becker, M. (2001). Managing dispersed knowledge: Organizational problems, managerial strategies, and their effectiveness. *Journal of Management Studies, 38*(7), 1037-1051.

Benbasat, I., & Zmud, R. (2003). The identity crisis within the IS discipline: Defining and communicating the discipline's core properties. *MIS Quarterly, 27*(2), 183-194.

Berger, P., & Luckmann, T. (1967). *The social construction of reality: A treatise in the sociology of knowledge.* Garden City, NJ: Doubleday.

Bernecker, S., & Dretske, F. (Eds.). (2000). *Knowledge: Readings in contemporary epistemology.* New York: Oxford University Press.

Bieber, M., Engelbart, D., Furuta, R., Roxanne-Hiltz, S., Noll, J., Preece, J., Stohr, E., Turoff, M., & Van de Walle, B. (2002). Towards virtual community knowledge evolution. *Journal of Management Information Systems, 18*(4), 11-36.

Bird, A. (1994). Careers as repositories of knowledge: A new perspective on boundaryless careers. *Journal of Organizational Behavior, 15*(4), 325-345.

Bishop, C. (1995). *Neural networks for pattern recognition.* New York: Oxford University Press.

Blackler, F. (1995). Knowledge, knowledge work, and organizations: An overview and interpretation. *Organization Studies, 16*(6), 1021-1046.

Bock, G., & Kim, Y. (2002). Breaking the myths of rewards: An exploratory study of attitudes about knowledge sharing. *Information Resource Management Journal, 15*(2), 14-21.

Boden, M. (1994). Precis of the creative mind: Myths and mechanisms. *Behavioral and Brain Sciences, 17*(3), 519-570.

Boland, R.J., & Tenkasi, R.V. (1995). Perspective making and perspective taking in communities of knowledge. *Organization Science, 6*(4), 350-372.

Bonner, S., Libby, R., & Nelson, M. (1997). Audit category knowledge as a precondition to learning from experience. *Accounting, Organizations, and Society, 22*(5), 387-410.

Borgman, A. (1993). *Crossing the postmodern divide.* Chicago: University of Chicago Press.

Bornemann, M., & Sammer, M. (2003). Assessment methodology to prioritize knowledge management related activities to support organizational excellence. *Measuring Business Excellence, 7*(2), 21-29.

Bottazzi, G., Dosi, G., & Rocchetti, G. (2001). Modes of knowledge accumulation entry regimes, and patterns of industrial evolution. *Industrial and Corporate Change, 10*(3), 609-638.

Boukendour, S., & Brissaud, D. (2005). A phenomenological taxonomy for systematizing knowledge on nonconformaces. *The Quality Management Journal, 12*(2), 25-33.

Bouvier, A. (2002). An epistemological plea for methodological individualism and rational choice theory in cognitive rhetoric. *Philosophy of the Social Sciences, 32*(1), 51-70.

Braddon-Mitchell, D., & Jackson, F. (1996). *The philosophy of mind and cognition.* London: Blackwell.

Bradley, R. (1999). Explorations in economic methodology: From Lakatos to empirical philosophy of science. *The British Journal for the Philosophy of Science, 50*(2), 316-318.

Braganza, A., Edwards, C., & Lambert, R. (1999). A taxonomy of knowledge projects to underpin organizational innovation and competitiveness. *Knowledge and Process Management, 6*(2), 83-90.

Brazelton, J., & Gorry, G. (2003). Creating a knowledge-sharing community: If you build it, will they come? *Communications of the ACM, 46*(2), 23-25.

Brief, J. (1983). *Beyond Piaget: A philosophical psychology.* New York: Teachers College Press.

Brown, J., & Duguid, P. (1998). Organizing knowledge. *California Management Review, 40*(3), 90-111.

Brown, J., & Duguid, P. (2001). Knowledge and organization: A social-practice perspective. *Organization Science, 12*(2), 198-213.

Brown, J.S., & Duguid, P. (1991). Organizational learning and communities of practice: Toward a unified view of working, learning, and innovation. *Organization Science, 2*(1), 40-57.

Browne, A. (1997). *Neural network perspectives on cognition and adaptive robotics.* Bristol, UK: Institute of Physics Publishing.

Buckman, R. (2004). *Building a knowledge-driven organization.* New York: McGraw-Hill.

Buniyamin, N., & Barber, K. (2004). The intranet: A platform for knowledge management system based on knowledge mapping. *International Journal of Technology Management, 28*(7/8), 729-746.

Burke, J. (1999). *The knowledge web.* New York: Simon & Schuster.

Burke, P. (2000). *A social history of knowledge: From Gutenberg to Diderot.* Cambridge, UK: Polity Press.

Butos, W. (2003). Knowledge questions: Hayek, Keynes and beyond. *Review of Austrian Economics, 16*(4), 291-306.

Cai, Y., & Terrill, J. (2006). Visual analysis of human dynamics: An introduction to the special issue. *Information Visualization, 5*(3), 235-236.

Calude, C. (2002). Incompleteness, complexity, randomness, and beyond. *Minds and Machines, 12*(4), 503-517.

Camerer, C., Lowenstein, G., & Weber, M. (1989). The curse of knowledge in economic settings: An experimental analysis. *Journal of Political Economy, 97*(3), 1232-1254.

Cancho, R. (2005). The variation of Zipf's law in human language. *European Physical Journal, 44*(2), 249-257.

Capra, F. (2002). *The hidden connections: Integrating the biological, cognitive, and social dimensions of life into a science of sustainability.* New York: Doubleday.

Carayannis, E., & Campbell, D. (Eds.). (2005). *Knowledge creation diffusion and use in innovation networks and knowledge clusters.* Westport, CT: Praeger.

Carile, P. (2004). Transferring, translating and transforming: An integrative framework for managing knowledge across boundaries. *Organization Science, 15*(5), 555-568.

Carlile, P. (2002). A pragmatic view of knowledge and boundaries: Boundary objects in new product development. *Organization Science, 13*(4), 442-455.

Carlucci, D., Marr, B., & Schiuma, G. (2004). The knowledge value-chain: How intellectual capital impacts on business performance. *International Journal of Technology Management, 27*(6/7), 575-590.

Carnap, R. (1983). *The logical structure of the world and pseudo problems of philosophy* (reprint ed.). Sacramento, CA: University of California Press.

Carnap, R. (1995a). *An introduction to the philosophy of science.* New York: Dover.

Carnap, R. (1995b). *The unity of science.* Bristol, UK: Thoemmes Press.

Carrillo, F. (Ed.). (2005). *Knowledge cities: Approaches, experiences, and perspectives.* Burlington, MA: Butterworth-Heinemann.

Carrillo, J., & Gaimon, C. (2004). Managing knowledge-based resource capabilities under uncertainty. *Management Science, 50*(11), 1504-1518.

Cartwright, N. (2003). *The dappled world: A study of the boundaries of science.* New York: Cambridge University Press.

Cassirer, E. (1965). *Philosophy of symbolic forms: The phenomenology of knowledge.* New Haven, CT: Yale University Press.

Chafetz, J. (1978). *Primer on the construction and testing of theories in sociology.* Itasca, IL: F. E. Peacock.

Chakrabarti, A.K., Dror, I., & Eakabuse, N. (1993). Interorganizational transfer of knowledge: An analysis of patent citations of a defense firm. *IEEE Transactions on Engineering Management, 40*(1), 91-94.

Chatzel, J. (2003). *Knowledge capital: How knowledge-based enterprises really get built.* New York: Oxford University Press.

Chatzkel, J. (2003). *Knowledge capital.* New York: Oxford University Press.

Chauvel, D., & Despres, C. (2004). Organizational logic in the new age of business: The case example of knowledge management at Valtech. *International Journal of Technology Management, 27*(6/7), 611-627.

Chen, A., & Eddington, T. (2005). Assessing value in organizational knowledge creation: Considerations for knowledge workers. *MIS Quarterly, 29*(2), 279-310.

Churchland, P. (1988). *Neurophilosophy: Toward a unified science of the mind/brain.* Cambridge, MA: MIT Press.

Churchman, C.W. (1971). *The design of inquiry systems.* New York: Basic Books.

Clancey, W. (1997). *Situated cognition: On human knowledge and computer representations.* New York: Cambridge University Press.

Cohen, M.D. (1991). Individual learning and organizational routine: Emerging connections. *Organization Science, 2,* 135-139.

Cohen, W., & Levinthal, D. (1990). Absorptive capacity: A new perspective on learning and innovation. *Administrative Science Quarterly, 35*(2), 128-152.

Cole, P. (2002). *Access to philosophy: The theory of knowledge.* London: Hodder & Staughton Educational Division.

Coley, J., Hayes, B., Lawson, C., & Moloney, M. (2004). Knowledge, expectations, and inductive reasoning with conceptual hierarchies. *Cognition, 90*(2), 217-253.

Collins, H. (1990). *Artificial experts: Social knowledge and intelligent machines.* Cambridge, MA: MIT Press.

Collins, H. (1993). The structure of knowledge. *Social Research, 60*(2), 95-116.

Connant, J., Mokwa, M., & Varadarajan, D. (1990). Strategic types, distinctive marketing competencies, and organizational performance: A multiple measures-based study. *Strategic Management Journal, 11*(5), 365-383.

Console, L. et al. (2002). Local reasoning and knowledge compilation for efficient temporal abduction. *IEEE Transactions on Knowledge and Data Engineering, 14*(6), 1230-1248.

Cooper, G., & Sweller, J. (1987). Effects of schema acquisition and rule automation on mathematical problem-solving transfer. *Journal of Educational Psychology, 79*(4), 347-362.

Cooper, J. (1998). *Reason and emotion.* Princeton, NJ: Princeton University Press.

Cooper, J. (2004). *Knowledge, nature, and the good: Essays on ancient philosophy.* Princeton, NJ: Princeton University Press.

Cross, B., Parker, A., & Sasson, L. (Eds.). (2003). *Networks in the knowledge economy.* New York: Oxford University Press.

Cross, C. (2001b). A theorem concerning syntactical treatments of nonidealized beliefs. *Synthese, 129,* 335-341.

Cytowic, R. (2002). *Synaesthesia: A union of the senses* (2nd ed.). New York: Springer-Verlag.

Cziko, G. (1995). *Without miracles: Universal selection theory and the second Darwinian revolution.* Cambridge, MA: MIT Press.

D'Adderio, L. (2003). Configuring software, reconfiguring memories: The influence of integrated systems on the reproduction of knowledge and routines. *Industrial and Corporate Change, 12*(2), 321-350.

Daft, R.L., & Weick, K.E. (1984). Toward a model of organizations as interpretation systems. *Academy of Management Review, 9*(2), 285-295.

Daft, R.L., & Wiginton, J. (1979). Language and organization. *Academy of Management Review, 4*(2), 179-191.

Damodaran, L., & Olphert, W. (2000). Barriers and facilitators to the use of knowledge management systems. *Behavior and Information Technology, 19*(6), 405-413.

Davenport, T., & Harris, J. (2005). Automated decision making comes of age. *Sloan Management Review, 46*(4), 83-89.

Davenport, T.H., Delong, D.W., & Beers, M.C. (1998). Successful knowledge management projects. *Sloan Management Review, 39*(2), 43-57.

Davenport, T.H., Thomas, R.J., & Cantrell, S. (2002). The mysterious art and science of knowledge-worker performance. *Sloan Management Review, 44*(1), 23-30.

Davenport, T.H., Thomas, R.J., & Desouza, K.C. (2003). Reusing intellectual assets. *Industrial Management, 45*(3), 12-17.

de Laplante, K. (1999). Certainty and domain-independence in the sciences of complexity: A critique of James Franklin's account of formal science. *Studies in History and Philosophy of Science, 30*(4), 699-720.

Derrida, J. (2001). *Margins of philosophy.* Chicago: University of Chicago Press.

DeSouza, K. (2003). Knowledge management barriers: Why technology imperatives seldom works. *Business Horizons, 46*(1), 25-29.

Desouza, K.C. (2002). *Managing knowledge with artificial intelligence.* Westport, CT: Quorum Books.

Desouza, K.C., & Evaristo, J.R. (2003). Global knowledge management strategies. *European Management Journal, 21*(1), 62-67.

Desouza, K.C., & Hensgen, T. (2002). On "information" in organizations: An emergent information theory and semiotic framework. *Emergence: A Journal of Complexity Issues in Organizations and Management, 4*(3), 95-114.

Deutsch, D. (1997). *The fabric of reality.* New York: Allen Lane.

Devaraj, S., & Kohli, R. (2003). Performance impacts of information technology: Is actual usage the missing link? *Management Science, 49*(3), 273-289.

Dienes, Z., & Perner, J. (1999). A theory of implicit and explicit knowledge. *Behavioral and Brain Sciences, 22*(5), 162-231.

Dierkes, M., Berthoin, A., Child, J., & Nonaka, I. (Eds.). (2001). *Handbook of organizational learning.* Oxford, UK: Oxford University Press.

Dretske, F. (1981). The pragmatic dimension of knowledge. *Philosophical Studies, 40*(3), 363-378.

Dreyfus, H., & Dreyfus, S. (1986). *Mind over machine: The power of human intuition and expertise in the era of the computer.* New York: The Free Press.

DuPuy, J.-P. (1994). *The mechanization of the mind.* Princeton, NJ: Princeton University Press.

Eco, U. (1976). *A theory of semiotics.* Bloomington: Indiana University Press.

Einstein, A. (1955). *The meaning of relativity.* Princeton, NJ: Princeton University Press.

Elsbach, K., Barr, P., & Hargadon, A. (2005). Identifying situated cognition in organizations. *Organization Science, 16*(4), 422-433.

Elster, J. (1979). *Ulysses and the sirens: Studies in rationality and irrationality.* Cambridge, UK: Cambridge University Press.

Elster, J. (1999). *Alchemies of the mind: Rationality and the emotions.* New York: Cambridge University Press.

Elster, J. (2000). *Ulysses unbound.* New York: Cambridge University Press.

Eppler, M. (2001). Making knowledge visible through intranet knowledge maps: Concepts, elements, and cases. *Proceedings of the 34th Annual Hawaii International Conference on Systems Sciences,* Maui, HI.

Etzioni, A. (1991). Embedding decision-analytic control in a learning architecture. *Artificial Intelligence, 49*(3), 129-159.

Fahey, L., & Prusak, L. (1998). The eleven deadliest sins of knowledge management. *California Management Review, 40*(3), 265-279.

Fauctioner, G., & Turner, M. (2002). *The way we think: Conceptual blending and the mind's hidden complexities.* New York: Basic Books.

Fausett, L. (1994). *Fundamentals of neural networks.* Upper Saddle River, NJ: Prentice Hall.

Feng, K., Chen, E., & Liou, W. (2005). Implementation of knowledge management systems and firm performance: An empirical investigation. *Journal of Computer Information Systems, 45*(2), 92-104.

Feyerabend, P. (1995). *Realism, rationalism, and scientific method: Philosophical papers, volume I.* New York: Cambridge University Press.

Fikes, R., & Kehler, T. (1985). The role of frame-based representation in reasoning. *Communications of the ACM, 28*(7), 901-906.

Firat, A., Madnick, S., & Grosof, B. (2002, October). *Knowledge integration to overcome ontological heterogeneity: Challenges from financial information systems.* Unpublished Manuscript, MIT Sloan School of Management, USA.

Firestone, J., & McElroy, M. (2003). *Key issues in the new knowledge management.* Burlington, MA: Butterworth-Heinemann.

Fitzgerald, J., & Dennis, A. (1995). *Business data communications and networking* (5th ed.). New York: John Wiley & Sons.

Fleck, L. (1979). *Genesis and development of a scientific fact.* Chicago: University of Chicago Press.

Forster, M. (2004). *Wittgenstein on the arbitrariness of grammar.* Princeton, NJ: Princeton University Press.

Freer, S. (2003). *Linnaeus' philosophia botanica.* New York: Oxford University Press.

Frisina, W. (2002). *The unity of knowledge and action: Toward a nonrepresentational theory of knowledge.* Albany: State University of New York Press.

Fuller, S., Revere, D., Bugni, P., & Martin, G. (2004). A knowledge base system to enhance scientific discovery: Telemakus. *Biomedical Digital Libraries, 2.*

Galbraith, J.R., & Lawler, E.E. (1993). *Organizing for the future: The new logic for managing complex organizations.* San Francisco: Jossey-Bass.

Gallagher, S., & Hazlett, S.A. (2000). Using the knowledge management maturity model (KM3) as an evaluation tool. *Proceedings of the BPRC Conference on Knowledge Management: Concepts and Controversies,* Coventry, UK.

Gardner, H. (1983). *Frames of minds: The theory of multiple intelligences.* New York: Basic Books.

Gates, B., & Hemingway, C. (1999). *Business at the speed of thought: Using a digital nervous system.* New York: Warner Books.

Geisler, E. (1993, November). How professionals think: Explorations into cognitive patterns of knowledge processing. *Proceedings of the Southern Management Association.*

Geisler, E. (1995a). Measuring the unquantifiable: Issues in the use of indicators in unstructured phenomena. *International Journal of Operations and Quantitative Management, 1*(2), 145-161.

Geisler, E. (1995b). Strategic perspectives of artificial management and organizational rationality. *Journal of Information Technology Management, 6*(4), 45-53.

Geisler, E. (1997). *Managing the aftermath of radical corporate change: Reengineering, restructuring, and reinvention.* Westport, CT: Quorum Books.

Geisler, E. (1999a). Harnessing the value of experience in the knowledge-driven firm. *Business Horizons, 42*(3), 18-26.

Geisler, E. (1999b). Mapping the knowledge-base of management of medical technology. *International Journal of Healthcare Technology and Management, 1*(1), 3-10.

Geisler, E. (2007). A typology of knowledge management: Strategic groups and role behavior in organizations. *Journal of Knowledge Management, 11*(1).

Geisler, E. (2007, forthcoming). *Knowledge and knowledge systems: Learning from the marvels of the mind.* Hershey, PA: Idea Group.

Geisler, E. (2007, under review). *Epistemetrics (the metrics of knowledge)—part one: What we measure.*

Geisler, E., & Wickramasinghe, N. (2007, under review). *Epistemetrics, scientometrics, and concepts of scientific inquiry: In search of convergence.*

Geisler, E., & Wickramasinghe, N. (2006, July). Epistemetrics: Conceptual domain and metrics of knowledge management. *Proceedings of the 3rd Knowledge Management Aston Conference,* Aston University, UK.

Gill, J. (2000). *The tacit mode: Michael Polanyi's postmodern philosophy (SUNY series in constructive postmodern thought).* Albany: State University of New York Press.

Given, L., & Olson, H. (2003). Knowledge organization in research: A conceptual model for organizing data. *Library Information Science Research, 25*(2), 157-176.

Godin, B. (2004). *The knowledge-based economy: Conceptual framework or buzzword.* Working Paper No. 24, Lavalle University, Canada.

Goldstone, R.L. (1998). Perceptual learning. *Annual Reviews of Psychology, 49,* 585-612.

Gottschalk, P. (2005). *Strategic knowledge management technology.* Hershey, PA: Idea Group.

Gould, S.J. (1991). Exaptation: A crucial tool for an evolutionary psychology. *Journal of Social Issues, 47,* 43-65.

Graff, T., & Jones, T. (2003). *Introduction to knowledge management.* Burlington, MA: Butterworth-Heinemann.

Grant, R.M. (1996). Toward a knowledge-based theory of the firm. *Strategic Management Journal, 17*(Winter Special Issue), 109-122.

Grantham, T. (2000). Evolutionary epistemology, social epistemology, and the Demic structure of science. *Biology and Philosophy, 15*(3), 443-463.

Grayson, J., & O'Dell, C. (1998). *If only we knew what we know: The transfer of internal knowledge and best practice.* New York: The Free Press.

Greeno, J. (1998). The situativity of knowing, learning, and research. *American Psychologist, 53*(1), 5-26.

Gupta, S., Bhatanagar, V., & Wasan, S. (2005). Architecture for knowledge discovery and knowledge management. *Knowledge and Information Systems, 7*(3), 310-319.

Gurian, M. (2004). *What could he be thinking?: How a man's mind really works.* New York: St. Martin's Griffin.

Guth, A. (1997). *The inflationary universe.* Reading, MA: Perseus Press.

Habermas, J. (1985). *The theory of communicative action, volume 1: Reason and the rationalization of society.* Boston: Beacon Press.

Handzic, M. (2005). *Knowledge management: Through the technology glass.* Hackensack, NJ: World Scientific.

Hansen, M. (1999). The search-transfer problem: The role of weak ties in sharing knowledge across organizational subunits. *Administrative Science Quarterly, 44,* 83-111.

Hansen, M.T., Nohira, N., & Tierney, T. (1999). What's your strategy for managing knowledge? *Harvard Business Review, 77*(2), 106-116.

Harada, T. (2003). Three steps in knowledge communication: The emergence of knowledge transformers. *Research Policy, 32*(10), 1737-1751.

Hargadon, A., & Fanelli, A. (2002). Action and possibility: Reconciling dual perspectives of knowledge in organizations. *Organization Science, 13*(3), 290-302.

Hargitta, I., & Hargitta, M. (2006). *Candid science 6: More conversations with famous scientists.* London: Imperial College Press.

Harnad, S. (1991). Post-Gutenberg galaxy: The fourth revolution in the means of production of knowledge. *Public-Access Computer Systems Review, 2*(1), 39-53.

Harrigan, K. (1985). An application of clustering for strategic group analysis. *Strategic Management Journal, 6*(1), 55-73.

Hatten, K., & Rosenthal, S. (2001). *Reaching for the knowledge edge: How the knowing corporation seeks shares and uses knowledge for strategic advantage.* New York: AMACOM.

Hawking, S. (2001). *The universe in a nutshell.* New York: Bantam Books.

Haykin, S. (1998). *Neural networks: A comprehensive foundation.* Upper Saddle River, NJ: Prentice Hall.

Heeseok, L., & Byounggu, C. (2003). Knowledge management enablers, processes, and organizational performance: An integrative view and empirical examination. *Journal of Management Information Systems, 20*(1), 179-228.

Heidegger, M. (1969). *Discourse on thinking.* New York: Harper-Collins-Perennial.

Heidegger, M. (2000). *Being and time.* London: Blackwell.

Heinrichs, J., & Lim, J. (2005). Model for organizational knowledge creation and strategic use of information. *Journal of the American Society for Information Science and Technology, 56*(6), 620-631.

Heyes, C., & Plotkin, H. (1989). Replicators and interactors in cultural revolution. In M. Ruse (Ed.), *What the philosophy of biology is.* Dordrecht, The Netherlands: Kluwer Academic.

Hiebeler, R. (1997). Benchmarking knowledge. *Executive Excellence, 14*(1), 11-12.

Hinds, P. (1999). The curse of expertise: The effects of expertise and de-biasing methods on predictions of novice performance. *Journal of Experimental Psychology, 5*(3), 205-221.

Hirschfield, L., & Gelman, S. (Eds.). (1994). *Mapping the mind: Domain specificity in cognition and culture.* New York: Cambridge University Press.

Hirschheim, R. (1985). Information systems epistemology: An historical perspective. In E. Mumford, R. Hirschheim, G. Fitzgerald, & T. Wood-Harper (Eds.), *Research methods in information systems* (p. 1335). Amsterdam: North Holland.

Hirschheim, R., Klein, H.K., & Lyytinen, K. (1995). *Information systems development and data modeling: Conceptual and philosophical foundations.* Cambridge: Cambridge University Press.

Hirst, G. (1991). Existence assumptions in knowledge representation. *Artificial Intelligence, 49*(3), 199-242.

Holmes, N. (2004). Data and information as property. *Computer, 37*(5), 90-92.

Holsapple, C. (2002). *Handbook of knowledge management: Knowledge matters, volume 1.* New York: Springer-Verlag.

Holsapple, C., & Jones, K. (2004). Exploring primary activities of the knowledge chain. *Knowledge and Process Management, 11*(3), 155-174.

Holsapple, C., & Jones, K. (2005). Exploring secondary activities of the knowledge chain. *Knowledge and Process Management, 12*(1), 3-31.

Horodecki, R., Horodecki, M., & Horodecki, P. (2004). Quantum information isomorphism: Beyond the dilemma of the Scylla of antology and the Charybdis of instrumentalism. *IBM Journal of Research & Development, 48*(1), 139-147.

Housel, T., & Bell, A. (2001). *Measuring and managing knowledge.* New York: McGraw-Hill.

Huber, G. (1982). Organizations information systems: Determinants of their performance and behavior. *Management Science, 28*(2), 138-155.

Huber, G. (2001). Transfer of knowledge in knowledge management systems: Unexplored issues and suggested studies. *European Journal of Information Systems, 10,* 72-79.

Huff, A. (2000). Changes in organizational knowledge production. *Academy of Management Review, 25*(2), 288-293.

Huhns, M., & Glasser, L. (Eds.). (1989). *Readings in distributed artificial intelligence.* Menlo Park, CA: Morgan Kaufman.

Humphrey, N. (1999b). *Leaps of faith: Science, miracles, and the search for supernatural consolation.* New York: Copernicus Books.

Jashapara, A. (2005). The emerging discourse of knowledge management: A new dawn for information science research. *Journal of Information Science, 31*(2), 136-148.

Johnson, S. (2004). *Mind wide open: Your brain and the neuroscience of everyday life.* New York: Charles Scribner's Sons.

Kankanhalli, A., & Tan, B. (2004, January). A review for metrics for knowledge management systems and knowledge management initiatives. *Proceedings of the 37th Hawaii International Conference on Systems Science.*

Keil, F. (1989). *Concepts, kinds, and cognitive development.* Cambridge, MA: MIT Press.

Kelloway, K., & Barling, J. (1990). Item content versus wording: Disentangling role conflict and role ambiguity. *Journal of Applied Psychology, 75*(6), 738-745.

Kelly, K. (1996). *The logic of reliable inquiry.* New York: Oxford University Press.

Kermally, S. (2002). *Effective knowledge management: A best practice approach.* New York: John Wiley & Sons.

Killman, R.H. (1979). On integrating knowledge utilization with knowledge development: The philosophy behind the MAPS design psychology. *Academy of Management Journal, 4*(3), 417-426.

Kitcher, P. (2004). Evolutionary theory and the social uses of biology. *Biology and Philosophy, 19*(1), 1-15.

Klobas, J., & McGill, T. (1995). Identification of technological gatekeepers in the information technology profession. *Journal of the American Society for Information Science, 46*(8), 581-590.

Knorr-Cetina, K. (1999). *Epistemic cultures: How the sciences make knowledge.* Cambridge, MA: Harvard University Press.

Koenig, M., & Srikantaiah, K. (Eds.). (2004). *Knowledge management: Lessons learned.* Medford, NJ: Information Today.

Kogut, B., & Zander, U. (1992). Knowledge of the firm, combinative capabilities, and the replication of technology. *Organization Science, 3*(3), 383-397.

Koons, R. (1994). A new solution to the Sorites problem. *Minds and Machines, 103*(412), 439-449.

Kostoff, R., & Geisler, E. (1999). Strategic management and implementation of textual data mining in government organizations. *Technology Analysis & Strategic Management, 13*(2), 195-206.

Krippendorff, K. (1986). *Information theory: Structural models for qualitative data.* Beverly Hills, CA: Sage.

Kuhn, T. (1970). *The structure of scientific revolutions* (2nd ed.). Chicago: The University of Chicago Press.

Kuhn, T., & Gorman, S. (2003). The emergence of homogeneity and heterogeneity in knowledge structures during a planned organizational change. *Communication Management, 70*(3), 198-229.

Kurtz, C., & Snowden, D. (2003). The new dynamics of strategy: Sense-making in a complex and complicated world. *IBM Systems Journal, 42*(3), 462-483.

Kurzweil, R. (2000). *The age of spiritual machines: When computers exceed human intelligence.* New York: Penguin Books.

Lakatos, I. (1994). *The methodology of scientific research programs.* New York: Cambridge University Press.

Lakatos, I., & Musgrave, A. (Eds.). (1970). *Criticism and the growth of knowledge: Volume 4 proceedings of the international colloquium in the philosophy of science, London, 1965.* New York: Cambridge University Press.

Larichev, O. (2002). Close imitation of expert knowledge: The problem and methods. *International Journal of Information Technology & Decision Making, 1*(1), 27-42.

Latour, B. (1999). *Pandora's hope: Essays on the reality of science studies.* Cambridge, MA: Harvard University Press.

Lau, C. (1992). *Neural networks: Theoretical foundations and analysis.* New York: IEEE.

Lave, J. (1988). *Cognition in practice: Mind, mathematics and culture in everyday life.* New York: Cambridge University Press.

Lee, J.H., & Kim, Y.G. (2001). A stage model of organizational knowledge management: A latent content analysis. *Expert Systems with Applications, 20,* 299-311.

Lee, K., Lee, S., & Kang, W. (2005). KMPI: Measuring knowledge management performance. *Information & Management, 42*(3), 469-479.

Leonard-Barton, D. (1995). *Wellsprings of knowledge.* Boston: Harvard Business School Press.

Leplin, J. (2001). Lakatos' epistemic aspirations. *Theories, 16*(3), 481-498.

Lesser, E., & Prusak, L. (Eds.). (2003). *Creating value with knowledge.* New York: Oxford University Press.

Levi, I. (1997). *The covenant of reason: Rationality and the commitments of thought.* New York: Cambridge University Press.

Lewis, D. (1996). Elusive knowledge. *AustralAsian Journal of Philosophy, 74*(1), 549-567.

Lewis, K. (2004). Knowledge and performance in knowledge-worker teams: A longitudinal study of transactive memory systems. *Management Science, 50*(11), 1519-1534.

Leydesdorff, L. (1994). Exchange on the cognitive dimension as a problem for empirical research in science studies. *Social Epistemology, 8*(2), 91-107.

Leydesdorff, L. (2001a). *The challenge of scientometrics.* New York: Universal.

Leyesdorff, L., Dolfsma, W., & Van der Panne, G. (2004, June 10-12). Measuring the knowledge base of an economy in terms of triple-helix relations among technology, organization, and territory. *Proceedings of the 10th Annual Conference of the International Joseph A. Schumpeter Society,* Milan, Italy.

Liang, T. (1998). General information theory: Some macroscopic dynamics of the human thinking system. *Information Processing & Management, 34*(2/3), 275-290.

Liben, L. (1983). *Piaget and the foundations of knowledge.* Mahwah, NJ: Lawrence Erlbaum.

Liebowitz, J. (1999). *Knowledge management handbook.* Boca Raton, FL: CRC Press.

Liebowitz, J., & Wilcox, L.C. (1997). *Knowledge management and its integrative elements.* Boston: CRC Press.

Lippert, S., & Forman, H. (2005). Utilization of information technology: Examining cognitive and experiential factors of post-adoption behavior. *IEEE Transactions on Engineering Management, 52*(3), 363-381.

Lynn, G., Reilly, R., & Akgun, A. (2000). Knowledge management in new product teams: Practices and outcomes. *IEEE Transactions on Engineering Management, 47*(2), 221-231.

Lyons, M. (2005). Knowledge and the modeling of complex systems. *Futures, 37*(7), 711-723.

Maasen, S., & Weingart, P. (2001). *Metaphors and the dynamics of knowledge.* London: Routledge.

MacKenzie, D. (1995). *Knowing machines.* Cambridge, MA: MIT Press.

Malhotra, Y. (2001). Expert systems for knowledge management: Crossing the chasm between information processing and sense making. *Expert Systems with Applications, 20*(March), 7-16.

Markus, M.L. (2001). Toward a theory of knowledge reuse: Types of knowledge reuse situations and factors in reuse success. *Journal of Management Information Systems, 18*(1), 57-93.

Marti, J. (2004). Strategic Knowledge Benchmarking System (SKBS): A knowledge-based strategic management information system for firms. *Journal of Knowledge Management, 8*(6), 31-49.

Masterton, S., & Watt, S. (2000). Oracles, bards, and village gossips, or social roles and meta knowledge management. *Information Systems Frontiers, 2*(3-4), 299-315.

McCllelan, J., & Dorn, H. (1999). *Science and technology in world history.* Baltimore: Johns Hopkins University Press.

McDermott, R. (1999). Why information technology inspired but cannot deliver knowledge management. *California Management Review, 41*(4), 103-117.

McDowell, J. (1995). *Mind and world.* Cambridge, MA: Harvard University Press.

McFadyen, A., & Cannella, A. (2004). Social capital and knowledge creation: Diminishing returns of the number and strength of exchange relationships. *Academy of Management Journal, 47*(5), 735-746.

McInerney, C. (2002). Knowledge management and the dynamic nature of knowledge. *Journal of the American Society for Information Science and Technology, 53*(12), 1009-1018.

McTaggart, J. (1908). The unreality of time. *Mind, 17*(3), 456-473.

Merleau-Ponty, M. (1974). *Adventures of the dialectic.* London: Heinemann.

Merx-Chermin, M., & Nijhof, W. (2005). Factors influencing knowledge creation and innovation in an organization. *Journal of European Industrial Training, 29*(2/3), 135-149.

Miles, R., & Snow, C. (1978). *Organization strategy, structure, and process.* New York: McGraw-Hill.

Mills, E. (2002). Fallibility and the phenomenal Sorites. *Nous, 36*(3), 384-407.

Minsky, M. (1975). A framework for representing knowledge. In P. Winston (Ed.), *The psychology of computer vision* (pp. 211-277). New York: McGraw-Hill.

Minsky, M. (1985). *Society of mind.* New York: Simon and Schuster.

Misa, T. (2001). What are techno memes anyway? *American Scientist, 89*(1), 88-89.

Moenaert, R., & Souder, W. (1996). Context and antecedents of information utility at the R&D/marketing interface. *Management Science, 42*(11), 1592-1610.

Moore, S., Shiell, A., Hawe, P., & Haines, V. (2005). The privileging of communitarian ideas: Citation practices and the translation of social capital into public health research. *American Journal of Public Health, 95*(8), 1330-1337.

Moreland, R.L., Argote, L., & Krishnan, R. (1996). Socially shared cognition at work: Transitive memory and group performance. In J. Nye & A. Brower (Eds.), *What's social about social cognition? Research on socially shared cognition in small groups.* Thousand Oaks, CA: Sage.

Mortimer, M. (1999). *Learn Dewey Decimal Classification* (21st ed.). Lanham, MD: Scarecrow Press.

Moser, P., & Copan, P. (2003). *The rationality of theism.* New York: Routledge.

Mothe, J., & Foray, D. (Eds.). (2001). *Knowledge management in the innovation process.* Boston: Kluwer Academic.

Musallam, S. et al. (2004). Cognitive control signals for neural prosthetics. *Science, 305*(5681), 258-262.

Nelson, R. (1982). The role of knowledge in R&D efficiency. *The Quarterly Journal of Economics,* (August), 453-470.

Nelson, R., & Winter, R. (1982). *An evolutionary theory of economic change.* Cambridge, MA: Belknap.

Newell, A., & Rosenbloom, P. (1981). Mechanisms of skill acquisition and the law of practice. In J.R. Anderson (Ed.), *Cognitive skills and their acquisition* (pp. 1-56). Hillsdale, NJ: Lawrence Erlbaum.

Nicolas, R. (2004). Knowledge management impacts on decision making process. *Journal of Knowledge Management, 8*(1), 20-31.

Nicolini, D., Gherardi, S., & Yanow, D. (Eds.). (2003). *Knowing in organizations: A practice-based approach.* Armonk, NY: M.E. Sharpe.

Nielsen, B. (2005). The role of knowledge embeddedness in the creation of synergies in strategic alliances. *Journal of Business Research, 58*(9), 1194-1212.

Nonaka, I. (1991). The knowledge-creating company. *Harvard Business Review, 69*(6), 96-104.

Nonaka, I. (1994). A dynamic theory of organizational knowledge creation. *Organization Science, 5*(1), 14-37.

Nonaka, I., & Konno, N. (1998). The concept of 'Ba': Building a foundation for knowledge creation. *California Management Review, 40*(3), 40-54.

Nonaka, I., & Toyama, R. (2003). The knowledge-creating theory revisited: Knowledge creation as a synthesizing process. *Knowledge Management Research and Practice, 1*(1), 2-9.

Noonan, H. (1999). *Routledge philosophy guidebook to Hume on knowledge.* New York: Routledge.

Oakeley, H. (1940). Epistemology and the logical syntax of language. *Mind, 49*(3), 427-444.

O'Connor, G., Rice, M., Peters, L., & Veryzer, R. (2003). Managing interdisciplinary, longitudinal research teams: Extending grounded theory building methodologies. *Organization Science, 14*(4), 353-371.

O'Mara, M. (2004). *Cities of knowledge: Cold War science and the search for the next Silicon Valley.* Princeton, NJ: Princeton University Press.

Orlikowski, W. (2002). Knowing in practice: Enacting a collective capability in distributed organizations. *Organization Science, 13*(3), 249-273.

Ormiston, G., & Schrift, A. (Eds.). (1990). *The hermeneutic tradition.* Albany: State University of New York Press.

Palmquist, J., & Ketola, L. (1999). Turning data into knowledge: Database marketing steers marketers to the future. *Marketing Research, 11*(2), 28-32.

Pappas, G., & Swain, M. (Eds.). (1978). *Essays on knowledge and justification.* Ithaca, NY: Cornell University Press.

Parsons, J. (1996). An information model based on classification theory. *Management Science, 42*(10), 1437-1445.

Parsons, J., & Wand, Y. (1997). Choosing classes in conceptual modeling. *Communications of the ACM, 40*(6), 63-71.

Pavitt, K. (1984). Sectoral patterns of technical change: Towards a taxonomy and a theory. *Research Policy, 13*(6), 343-374.

Pemberton, J., Stonehouse, G., & Francis, M. (2002). Black and Decker—towards a knowledge-centric organization. *Knowledge and Process Management, 9*(3), 178-189.

Pereira, A., & Lungarzo, C. (2005). A framework for the computational approach to cellular metabolism supporting neuronal activity. *3*(3), 87-92.

Perez, M., Sanchez, A., Carnicer, P., & Jimenez, J. (2002). Knowledge tasks and teleworking: A taxonomy model of feasibility adoption. *Journal of Knowledge Management, 6*(3), 272-284.

Perry, W., & Moffat, J. (1997). Measuring the effects of knowledge in military campaigns. *Journal of the Operational Research Society, 48*(10), 965-973.

Pesic, P. (2001). *Labyrinth: A search for the hidden meaning of science.* Cambridge, MA: MIT Press.

Petrash, G. (1996). Dow's journey to a knowledge value management culture. *European Management Journal, 14*(4), 365-373.

Piel, G., & Bradford, P. (2001). *The age of science: What scientists learned in the twentieth century.* New York: Basic Books.

Pirolli, P., & Recker, M. (1994). Learning strategies and transfer in the domain of programming. *Cognition and Instruction, 12*(3), 235-275.

Pivcevic, E. (1979). Truth as structure. *Review of Metaphysics, 28*(2), 311-327.

Plantinga, A., & Wolterstoff, N. (1984). *Faith and rationality: Reason and belief in God.* South Bend, IN: University of Notre Dame Press.

Plotkin, H. (1997). *Darwin machines and the nature of knowledge.* Cambridge, MA: Harvard University Press.

Plotkin, H. (1998). *Evolution in mind: An introduction to evolutionary psychology.* Cambridge, MA: Harvard University Press.

Poh-Kam, W. (2000). Knowledge creation management: Issues and challenges. *Asia-Pacific Journal of Management, 17*(2), 193-200.

Pollock, J., & Cruz, J. (1999). *Contemporary theories of knowledge.* Lanham, MD: Rowman & Littlefield.

Ponzi, L. (2003). *The evolution and intellectual development of knowledge management.* Unpublished Dissertation, Long Island University, USA.

Popper, K. (1959). *The logic of scientific discovery.* New York: Basic Books.

Popper, K. (1992). *Conjectures and refutations: The growth of scientific knowledge* (5th ed.). New York: Routledge.

Potter, V. (Ed.). (1993). *On understanding understanding: A philosophy of knowledge* (2nd ed.). New York: Fordham University Press.

Powell, W., & Snellman, K. (2004). The knowledge economy. *Annual Review of Sociology, 30*(2), 199-220.

Prusak, L. (2001). Where did knowledge management come from? *IBM Systems Journal, 40*(4), 1002-1007.

Purvis, R., Sambamurthy, V., & Zmud, R. (2000). The development of knowledge embeddedness in case technologies within organizations. *IEEE Transactions in Engineering Management, 47*(2), 245-257.

Radnitzky, G., & Bartley, W.W. (Eds.). (1987). *Evolutionary epistemology, rationality and the sociology of knowledge.* LaSalle, IL: Open Court.

Raghu, T. (1991). Perceptions of role stress by boundary role persons: An empirical investigation. *Journal of Applied Behavioral Sciences, 27*(4), 490-515.

Ramaprasad, A., & Ambrose, P.J. (1999). The semiotics of knowledge management. *Proceedings of the Workshop on Information Technology and Systems,* Charlotte, NC.

Ramaprasad, A., & Rai, A. (1996). Envisioning management of information. *Omega, International Journal of Management Science, 24*(2), 179-193.

Ramirez, Y., & Nembhard, D. (2004). Measuring knowledge worker productivity: A taxonomy. *Journal of Intellectual Capital, 5*(4), 602-628.

Reber, A. (1997). *Implicit learning and tacit knowledge: An essay on the cognitive unconscious.* New York: Oxford University Press.

Redko, V. (2000). Evolution of cognition: Towards the theory of origin of human logic. *Foundations of Science, 5*(2), 323-338.

Reid, T. (1941). *Essays on the intellectual powers of man.* London: McMillan.

Rescher, N. (2003a). *Epistemology: An introduction to the theory of knowledge.* Albany: State University of New York Press.

Revilla, E., Sarkis, J., & Acosta, J. (2005). Towards a knowledge management and learning taxonomy for research joint ventures. *Technovation, 25*(11), 1307-1316.

Roberts, J. (2000). From know-how to show-how? Questioning the role of information and communication technologies in knowledge transfer. *Technology Analysis & Strategic Management, 12*(4), 429-443.

Rogers, E. (1995). *Diffusion of innovations* (4th ed.). New York: The Free Press.

Roos, J., & Von Krogh, G. (1995). *Organizational epistemology.* New York: MacMillan.

Rotenstreich, N. (1998). *Synthesis and intentional objectivity: On Kant and Husserl.* Dordrecht, The Netherlands: Kluwer Academic.

Rowley, J. (2003). Knowledge management: The new librarianship? From custodians of history to gatekeepers to the future. *Library Management, 24*(8/9), 433-440.

Ruggles, R. (1998). The state of the notion: Knowledge management in practice. *California Management Review, 40*(3), 80-89.

Russell, B. (1998). *The problems of philosophy.* New York: Oxford University Press.

Russell, S., & Wefald, E. (1991). Principles of metareasoning. *Artificial Intelligence, 49*(3), 361-395.

Ryu, C., Kim, Y., Chaudhury, A., & Rao, H. (2005). Knowledge acquisition via three learning processes in enterprise information portals: Learning-by-investment, learning-by-doing, and learning-from-others. *MIS Quarterly, 29*(2), 245-289.

Sabherwal, & Becerra-Fernandez, I. (2005). Integrating specific knowledge: Insights from the Kennedy Space Center. *IEEE Transactions in Engineering Management, 52*(3), 301-315.

Santosus, M. (2001). Thanks for the memories: Northrop Grumman KM Case. *CIO Magazine,* (September 1), 26-29.

Savitt, S. (2000). *Time's arrow today.* Cambridge, UK: Cambridge University Press.

Sawyer, J. (1992). Goal and process clarity: Specification of multiple constructs of role ambiguity and a structural equation model of their antecedents and consequences. *Journal of Applied Psychology, 77*(2), 130-142.

Schaffer, J. (2001). Knowledge, relevant alternatives and missed clues. *Analysis, 61*(2), 202-208.

Schlick, M. (1992). *General theory of knowledge.* Chicago: Open Court.

Schopenhauer, A. (1966). *The world as will and representation.* Mineola, NY: Dover.

Schultze, U., & Leidner, D. (2002). Studying knowledge management in information systems research: Discourses and theoretical assumptions. *MIS Quarterly, 26*(3), 213-242.

Schwanellugel, P., Martin, M., & Takahashi, T. (1999). The organization of verbs of knowing: Evidence of a cultural commonality and variation in theory of mind. *Memory & Cognition, 27*(5), 813-825.

Senge, P. (1990). The leader's new work: Building a learning organization. *Sloan Management Review, 32*(1), 7-23.

Shekar, B., & Natarjan, R. (2004). A framework for evaluating knowledge-based interestingness of association rules. *Fuzzy Optimization and Decision Making, 3*(2), 157-185.

Shermer, M. (2001). *The borderlands of science: Where sense meets nonsense.* New York: Oxford University Press.

Silva, F., & Augusti-Cullel, J. (2003). Issues on knowledge coordination. *Knowledge and Process Management, 10*(1), 37-59.

Simner, J., & Ward, J. (2006). Synaesthesia: The taste of words on the tip of the tongue. *Nature, 444*(7118), 438-451.

Sinclair, N. (2006). *Stealth KM: Winning knowledge management strategies for the public sector.* Burlington, MA: Butterworth-Heinemann.

Singley, M.K., & Anderson, J.R. (1989). *The transfer of cognitive skills.* Cambridge, MA: Harvard Press.

Skyrme, D. (1998). *Measuring the value of knowledge: Metrics for the knowledge-based business.* New York: Business Intelligence.

Smith, C. (2002). Social epistemology, contextualism and the division of labor. *Social Epistemology, 16*(1), 65-81.

Smith, K., Collins, C., & Clark, K. (2005). Existing knowledge, knowledge creation capability, and the rate of new product introduction in high technology firms. *Academy of Management Journal, 48*(2), 346-365.

Sorensen, C., & Kakihara, M. (2002). Knowledge discourses and interaction technology. *Proceedings of the 35th International Conference on System Sciences* (HICSS-35), Big Island, HI.

Southon, F.C.G., Todd, R.J., & Seneque, M. (2002). Knowledge management in three organizations: An exploratory study. *Journal of the American Society for Information Science and Technology, 53*(12), 1047-1059.

Srilatha, P., & Harigopal, K. (1985). Role conflict and role ambiguity: Certain antecedents and consequences. *ASCI Journal of Management, 15*(1), 77-109.

Stalnaker, R. (1984). *Inquiry.* Cambridge, MA: MIT Press.

Star, S. (1989). *Regions of the mind: Brain research and the quest for scientific certainty.* Stanford, CA: Stanford University Press.

Starosta, G. (2003). Scientific knowledge and political action: On the antinomies of Lukacs' thought in history of class consciousness. *Science and Society, 67*(1), 39-68.

Stedman, H. et al. (2004). Myosin cell mutation correlates with anatomical changes in the human lineage. *Nature, 428*(6981), 415-418.

Stenger, V. (2001). *Timeless reality.* Amherst, NY: Prometheus Books.

Stengers, I. (2000). *The invention of modern science.* Minneapolis: University of Minnesota Press.

Sterelny, K. (2001). *Dawkins vs. Gould.* Cambridge, UK: Totem Books.

Sterelny, K., & Griffiths, P. (1979). *Sex and death: An introduction to the philosophy of biology.* Chicago: The University of Chicago Press.

Sternberg, R. (1985). *Beyond IQ: A triarchic theory of human intelligence.* New York: Cambridge University Press.

Sternberg, R. (1997). *Thinking styles.* New York: Cambridge University Press.

Sternberg, R. et al. (2000). *Practical intelligence.* New York: Cambridge University Press.

Stewart, A. (1997). *Elements of knowledge: Pragmatism, logic, and inquiry* (revised ed.). Nashville, TN: Vanderbilt University Press.

Stewart, T. (1997). *Intellectual capital: The new wealth of organizations.* New York: Doubleday.

Stokes, D. (1997). *Pasteur's Quadrant: Basic science and technological innovation.* Washington, DC: The Brookings Institute.

Strassmann, P. (2005). How much is know-how worth? *Baseline,* (November), 28, 79.

Stump, D. (2003). Defending conventions as functionally a priori knowledge. *Philosophy of Science, 70*(5), 1149-1161.

Sullivan, H. (2000). *Value driven intellectual capital.* New York: John Wiley & Sons.

Svieby, K. (1997). *The new organizational wealth: Managing and measuring knowledge-based assets.* San Francisco: Berrett-Koehler.

Swanson, D. (1986). Undiscovered public knowledge. *Library Quarterly, 56*(2), 103-118.

Swap, W., Leonard, D., Shields, M., & Abrams, I. (2001). Using mentoring and storytelling to transfer knowledge in the workplace. *Journal of Management Information Systems, 18*(1), 95-114.

Szulanski, G., Cappetta, R., & Jensen, R. (2004). When and how trustworthiness matters: Knowledge transfer and the moderating effect of causal ambiguity. *Organization Science, 15*(5), 600-613.

Takeda, H. (1999, August 2). *Proceedings of the IJCAI-99 Workshop on Ontologies and Problem-Solving Methods,* Stockholm, Sweden.

Teece, D.J. (1998). Research directions for knowledge management. *California Management Review, 40*(3), 289-292.

Ter Hark, M. (2003). *Popper, Otto Selz and the rise of evolutionary epistemology.* New York: Cambridge University Press.

Thagard, P. (1992). *Conceptual revolutions.* Princeton, NJ: Princeton University Press.

Thompson, M., & Walsham, G. (2004). Placing knowledge management in context. *Journal of Management Studies, 41*(5), 725-741.

Thow-Yick, L. (2004). *Organizing around intelligence.* London: World Scientific.

Tidd, J. (2001). *From knowledge management to strategic competence.* London: World Scientific.

Tomei, L. (Ed.). (2005). *Taxonomy for the technology domain.* Hershey, PA: Idea Group.

Travis, C. (2004). The silence of the senses. *Mind, 113*(449), 57-94.

Tsoukas, H. (1996). The firm as a distributed knowledge system: A constructionist approach. *Strategic Management Journal, 17*(1), 11-25.

Turchin, V. (1977). *The phenomenon of science: A cybernetic approach to human evolution.* New York: Columbia University Press.

Turing, A. (1936). On computable numbers, with an application to the Entscheidungs problem. *Proceedings of the London Mathematical Society* (series 2, vol. 42, pp. 230-65). Reprinted in David, M. (Ed.). (1965). *The undecidable.* Hewlett, NY: Raven Press.

Tushman, M.L., & Nadler, D.A. (1978). Information processing as an integrating concept in organizational design. *Academy of Management Review, 3*(3), 613-624.

Vadeboncoer, I., & Markovits, H. (1999). The effect of instructions and information retrieval on accepting the premises in a conditional reasoning task. *Thinking and Reasoning, 5*(2), 97-113.

Van der Meer, E., & Hoffmann, J. (Eds.). (1987). *Knowledge-aided information processing.* New York: Elsevier Science.

Van Lehn, K. (Ed.). (1992). *Architectures for intelligence: The twenty second Carnegie Mellon symposium on cognition.* Mahwah, NJ: Lawrence Erlbaum.

VanLehn, K. (1989). *Mind bugs: The origins of procedural misconception.* Cambridge, MA: MIT Press.

VanLehn, K. (1996). Cognitive skill acquisition. *Annual Reviews of Psychology, 47*, 513-539.

Vasconcelos, A., Souza, P., & Tribolet, J. Information systems architectures: Representation, planning, and evaluation. *Journal of Systemics, Cybernetics and Informatics, 1*(6), 96-104.

Ven, J., & Chuang, C. (2005). The comparative study of information competencies: Using Bloom's taxonomy. *Journal of American Academy of Business, 7*(1), 136-143.

Vincenti, W. (1993). *What engineers know and how they know it.* Baltimore: Johns Hopkins University Press.

Vinge, V. (1993). *The technological singularity.* Retrieved from *http://www.kurzwelai.net/articles/artoo92.html?Printable=1*

Vogel, J. (1999). The new relevant alternatives theory. *Philosophical Perspectives, 13*(3), 155-180.

Von Krogh, G., Ichijo, K., & Nonaka, I. (2000). *Enabling knowledge creation.* New York: Oxford University Press.

Vroom, V. (1964). *Work and motivation.* New York: John Wiley & Sons.

Walsh, J.P., & Ungson, G.R. (1991). Organizational memory. *Academy of Management Review, 16*(1), 57-91.

Wang, C., & Ahmed, P. (2004). Leveraging knowledge in the innovation and learning process at GKN. *International Journal of Technology Management, 26*(6/7), 674-688.

Weick, K. (1995). *Sensemaking in organizations.* Thousand Oaks, CA: Sage.

Weick, K., Sutcliffe, K., & Obstfeld, D. (2005). Organizing and the process of sensemaking. *Organization Science, 16*(4), 409-421.

Wenger, E. (1998). *Communities of practice.* New York: Cambridge University Press.

White, D. (Ed.). (2002). *Knowledge mapping and management.* Hershey, PA: Idea Group.

Whittlesea, B. (1987). Preservation of specific experiences in the representation of general knowledge. *Journal of Experimental Psychology, Learning, Memory, and Cognition, 13*(1), 3-17.

Wiig, K. (1999). What future knowledge management users may expect. *Journal of Knowledge Management, 3*(2), 155-165.

Williamson, T. (2002). *Knowledge and its limits.* New York: Oxford University Press.

Wilson, R. (2003). *Four colors suffice: How the map problem was solved.* Princeton, NJ: Princeton University Press.

Wittgenstein, L. (2001). *Tractatus logico-philosophicus.* Paris: Edition Gallimard.

Woit, P. (2006). *Not even wrong: The failure of string theory and the search for unity in physical law.* New York: Basic Books.

Wolfram, S. (2002). *A new kind of science.* Champaign, IL: Wolfram Media.

Womack, J.P., Jones, D.T., & Ross, D. (1990). *The machine that changed the world.* New York: Macmillan.

Wong, K., & Aspinwall, E. (2005). Knowledge management: Case studies in SMEs and evaluation of an integrated approach. *Journal of Information and Knowledge Management, 4*(2), 95-111.

Zack, M. (2003). Rethinking the knowledge-based organization. *Sloan Management Review, 44*(4), 67-71.

Zahra, S., & Pearce, J. (1990). Research evidence on the Miles-Snow typology. *Journal of Management, 16*(4), 751-768.

Zellner, C., & Fornahl, D. (2002). Scientific knowledge and implications for its diffusion. *Journal of Knowledge Management, 6*(2), 190-199.

Zheng, W. (2005). A conceptualization of the relationship between organizational culture and knowledge management. *Journal of Information & Knowledge Management, 4*(2), 113-124.

Zmud, R. (1978). An empirical investigation of the dimensionality of the concept of information. *Decision Sciences, 9*(2), 187-195.

Zuboff, S., & Maxmin, J. (2002). *The support economy: Why corporations are failing individuals and the next episode of capitalization.* New York: Penguin Press.

About the Contributors

Eliezer (Elie) Geisler is distinguished university professor at the Stuart School of Business, Illinois Institute of Technology, and director of the IIT Center for the Management of Medical Technology. He holds a doctorate from the Kellogg School at Northwestern University. Dr. Geisler is the author of nearly 100 papers in the areas of technology and innovation management, the evaluation of R&D, science and technology, and the management of healthcare and medical technology. He is the author of several books, including: *Managing the Aftermath of Radical Corporate Change* (1997); *The Metrics of Science and Technology* (2000), also translated into Chinese; *Creating Value with Science and Technology* (2001); *Technology, Healthcare and Management in the Hospital of the Future* (2003); and *Installing and Managing Workable Knowledge Management Systems* (with Rubenstein, 2003). He consulted for major corporations and for many U.S. federal departments, such as Defense, Agriculture, Commerce, EPA, Energy, and NASA. Dr. Geisler is the co-founder of the annual conferences on the Hospital of the Future and for the Health Care Technology and Management Association, a joint venture of several universities in 10 countries. He serves on various editorial boards of major journals. His current research interests include knowledge management in general and in complex systems, such as healthcare organizations.

* * * * *

Gerald M. Hoffman is president of the Gerald Hoffman Company LLC. He is a consultant, an educator, and an author. His most recent industrial job was as executive vice president and general manager of Amoco Computer Services Company.

His field of expertise is the management of information technology. He has done pioneering work in helping organizations get value from their information technology investments, in identifying and mapping technology trends and their effects on organizations, and in developing business strategies to exploit these trends. Most recently he has investigated the causes of IT project failures and developed a methodology help insure IT project success.

Index

Single Journal Articles and Case Studies
Are Now Right at Your Fingertips!

Purchase any single journal article or teaching case for only $25.00!

IGI Global offers an extensive collection of research articles and teaching cases in both print and electronic formats. You will find over 1300 journal articles and more than 300 case studies on-line at **www.igi-pub.com/articles**. Individual journal articles and cases are available for only $25 each. A new feature of our website now allows you to search journal articles and case studies by category. To take advantage of this new feature, simply use the above link to search within these available categories.

We have provided free access to the table of contents for each journal. Once you locate the specific article needed, you can purchase it through our easy and secure site.

For more information, contact cust@igi-pub.com or 717-533-8845 ext.10

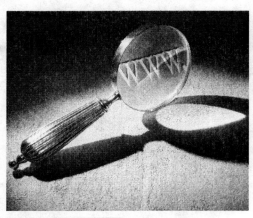

Databases, Data Mining & Data Warehousing

Distance Learning & Education

E-Commerce and E-Government

E-Government

Healthcare Information Systems

Human Side and Society Issues in IT

Information Technology Education

IT Business Value, Support and Solutions

IT Engineering, Modeling & Evaluation

Knowledge Management

Mobile Commerce and Telecommunications

Multimedia Networking

Virtual Organizations and Communities

Web Technologies and Applications

www.igi-pub.com